Editions SR / 24

EDITIONS SR

Volume 24

God and the Chip

Religion and the Culture of Technology

William A. Stahl

Published for the Canadian Corporation for Studies in Religion/
Corporation Canadienne des Sciences Religieuses
by Wilfrid Laurier University Press

This book has been published with the help of a grant from the Humanities and Social Sciences Federation of Canada, using funds provided by the Social Sciences and Humanities Research Council of Canada.

We acknowledge the support of the Canada Council for the Arts for our publishing program.

We acknowledge the financial support of the Government of Canada through the Book Publishing Industry Development Program for our publishing activities.

Canadian Cataloguing in Publication Data

Stahl, William A. (William Austin)
God and the chip : religion and the culture of technology

(Editions SR ; 24)
Includes bibliographical references and index.
ISBN 0-88920-321-0

1. Technology – Social aspects. 2. Technology – Moral and ethical aspects.
3. Computers – Social aspects. 4. Computers – Moral and ethical aspects.
I. Canadian Corporation for Studies in Religion. II. Title. III. Series.

T14.5.S72 1999 303.48′.3 C98-932486-9

Cover design by Leslie Macredie using an image by William Blake entitled *The Ancient of Days* (© Copyright The British Museum)

Printed in Canada

God and the Chip: Religion and the Culture of Technology has been produced from a manuscript supplied in camera-ready form by the author.

Order from:
Wilfrid Laurier University Press
Waterloo, Ontario, Canada N2L 3C5

Acknowledgments

I wish to thank the following publishers for permission to reprint and revise material previously published:

Dianoia, for "Technological Mysticism" Vol. 1, No. 1, Spring 1990: 1-17.

Science, Technology & Human Values, for "Venerating the Black Box: Magic in Media Discourse on Technology" Vol. 20, No. 2, Spring 1995: 234-58 ©1995 Sage Publications Inc. Reprinted by permission of Sage Publications Inc.

Material in Chapter Four from *Time* © 1977-1988, Time, Inc. All rights reserved. Used with permission.

Many people offered help and encouragement over the life of this project. I want to especially thank Reginald Bibby, Nancy Nason-Clark, Lori Walker, Peter Beyer, Lori Beaman, Edward Bailey and the people at the Denton Conferences on Implicit Religion, and, above all, my wife Ruth.

Table of Contents

Introduction

> We have a choice of what myths, what visions we will use to help us understand the physical world. We do not have a choice of understanding it without using any myths or visions at all. Again, we have a real choice between becoming aware of these myths and ignoring them. If we ignore them, we travel blindly inside myths and visions which are largely provided by other people. This makes it much harder to know where we are going.
>
> Mary Midgley
> *Science as Salvation*

It is perhaps the greatest irony of our times that in this technological age many, if not most, of our major problems are moral paradoxes. Computers and communications technology are in the forefront of changes which are transforming everyday life, yet we are critically short of ethical and intellectual resources with which to understand and confront these changes. Much of the language with which we discuss issues involving technology is ideological or mystifying. Indeed, much of it is magical and implicitly religious.

This should not, perhaps, be too surprising. Throughout history, most human cultures have surrounded technology with myth and ritual. To engage in creation was to participate in (or to encroach upon) the preserve of the gods, and weavers, potters, and especially smiths were commonly perceived as immersed in the sacred. "'To make' something means knowing the magic formula which will allow it to be invented or to 'make it appear' spontaneously," says Mircea Eliade (1978: 101-102). "In virtue of this, the artisan is a connoisseur of secrets, a magician; thus all crafts include some kind of initiation and are handed down by an occult tradition." Before the modern era, most peoples saw the material world as alive and they often personified nature as female. Those who would penetrate the mysteries of nature thus had to engage in propitiatory rites, particularly miners and metalworkers, who were often perceived as violating Mother Earth and who had to go through elaborate rituals of purification and sexual cleansing (Eliade, 1978; Merchant, 1980). This ambiguity towards technology and those who ply it was reflected in the

Greek myth of Hephaestus (the Roman Vulcan), the god of fire and the forge, who alone among the deities was ugly and lame (in a mythos that equated beauty with virtue and truth) and who was at one time cast out of Olympus (Hamilton, 1942).

Today, however, we claim to be different from our ancestors. We live in a secularized society. We are realists. Nature for us is dead matter, shaped by the impersonal forces described by science. Myth and ritual, to the extent they have any meaning at all, are matters of individual preference. We leave symbols to the poets and guide policy with fact and reason. Or so we claim.

The problem with these claims is that they strip technology and the world of the systems of meaning through which people have made sense of their lives and guided their actions. The language we use shapes and defines issues and problems. Part of the reason for our moral deafness is that the usual way of speaking about technology in our society is too restrictive. If today technology seems so paradoxical and morally baffling, it is in part because we no longer pay conscious attention to the kinds of symbols and rituals through which our ancestors regulated their interactions with nature and the material world. I say "pay conscious attention" because technology is still permeated with symbol and myth, but now they are implicit and hidden. So long as they remain hidden we give them power over us and we are subject to manipulation and self-deception. So ironically, the first step in the recovery of meaning is demythologization. Only when we have killed the idols, as Paul Ricoeur says (1970: 531), can we begin to make sense of the changes going on around us and in us, and begin to develop social ethics for an information age.

Historically, technology has not only been immersed in sacred myth and ritual, in many cultures certain technologies became central metaphors through which the theology, philosophy, literature and science of that society understood reality. J. David Bolter (1984) calls these "defining technologies." To the ancient Greeks, he claims, the spindle and the potter's wheel played such a role. In the Renaissance and early modern Europe, clocks were the defining technology, as exemplified by the clockwork imagery used by Descartes, Liebnitz, and Newton in describing nature. The nineteenth century used the steam engine as its metaphor of power and progress. Today, Bolter says, the computer is becoming a defining technology. Certainly, computers and other information technologies are on the cutting edge of social and economic changes. They are also changing the way we communicate, and therefore the way we think. There is a double aspect to these technologies. They are part of the content of discourse, they are also the medium of discourse—they are both what we publicly talk about and means by which we talk in public. Both aspects are shaped by cultural conventions and myths even as the technology reshapes them. For example, one of the means through which film and television communicates is camera angles (Tuchman, 1978; Hall, 1969). Technology creates the phenomenon of the camera angle, yet what these angles communicate are the meanings attributed to different perspectives which have developed in the West since the Renaissance (e.g., a shot looking up connotes authority, a close-up conveys intimacy). In doing this the technology both reproduces and reshapes cultural meanings at the same time. It is this double aspect of information and communication technology that makes them so important. Therefore I will use them as the focus for this work and the means by which we may understand technology more generally.

The study of technology is one of the preoccupations of the modern world. Unfortunately, this study is fragmented—historians of technology rarely cite sociologists, and vice versa (Staudenmaier, 1985, 1994), philosophers and theologians rarely consult social scientists, scientists and engineers, with a few notable exceptions, usually speak only to each other. In this work I try to overcome this fragmentation through an interdisciplinary approach. I will put the sociology of religion and science, technology, and society studies (STS) in conversation with each other, with comments from computer science, feminism, philosophy and theology.

I also deliberately try to blend critical analysis with a prophetic stance. I want to avoid either the uncritical celebration of the new technologies or a doomsaying jeremiad, two approaches which are far too common in public debates over technology. Too much of what passes for social ethics today is impassioned pleading deficient in social scientific analysis. Too much of what passes for social science today is devoid of moral perspective. Only by combining the best insights of theology and ethics with the most advanced tools of social science can there even be a hope of unraveling the moral paradoxes of modern technology.

One of the major problems of any interdisciplinary work is to translate the jargon of specialized subdisciplines so that others may understand it. I will try to keep jargon to a minimum. There are two central ideas which need some definition before we can begin, however, *implicit religion* from the sociology of religion and *constructivism* from STS.

What Is Implicit Religion?

Much of our language about technology is implicitly religious, a term that needs some explanation. I define implicit religion as those symbols and rituals directed to the numinous which are located outside formal religious organizations (e.g., churches) and which are often unrecognized, unacknowledged, or hidden. This definition does several things, which I will name now and develop more fully as we go along. Most obviously, implicit religion is implicit. It is not to be found, as such, in those institutions devoted to "explicit religion," although a number of studies (e.g., Bailey, 1983, 1990; Nesti, 1990; ter Borg, 1992) have shown that there are often close links between them. Nor is it usually open or acknowledged, and indeed people may be quite unaware of it as "religion." This is no more than the truism that "you can't expect a fish to discover water," or as Clifford Geertz said rather more profoundly, any religion clothes its symbols "with such an aura of facticity" that they "seem uniquely realistic" (Geertz, 1973: 90). In most cases, to the extent that an implicit religion is viable it will be "invisible."[1]

The most pointed question students of implicit religion have to answer is: "Is it *really* religion?" A flippant answer is to respond: "It depends on what you mean by religion." One thing that implicit religion is *not*, at least for me, is an excessively functionalist understanding of religion. Over the past seventy years a number of sociologists of religion have defined religion by what it does, that is, by the functions it performs in society. While this may have some merit, there is a danger in this approach that religious symbols and rituals can disappear into all their functions without remainder. Religion is more than what it does, it is, as Robert Bellah says, first and foremost "a set of symbolic forms and acts" which relate people to "the ultimate conditions of their existence" (Bellah, 1970: 21). We have to look to the substance of symbols and rituals to find how they ground meaning and identity in the transcendent.

Most students of implicit religion have followed Weber's dictum that it is only possible to define the essence of religion at the conclusion of empirical study (Weber, 1922/1963: 1). Thus Edward Bailey, for example, circles the question, describing implicit religion progressively as "commitments," "personal depths," "integrating foci," and "intensive concerns with extensive effects" (Bailey, 1983: 70-71). Similarly, Arnaldo Nesti traces its antecedents and central lines of research before declaring:

> Implicit religion is a request for meaning that originates in the subject's life-world, expressing itself by means of a complex system of symbols and practices. These in addition to a series of paths replete with meaning have the effect of reassuring the subjects themselves on the unconditional relevance of their existence here and now. (1990: 432)

---

[1] The phrase is, of course, Thomas Luckmann's (1967). While some might consider Luckmann to be the "founder" of the study of implicit religion, a review of the literature leads me to trace its main concepts to Durkheim. Although Luckmann's work contains many valuable insights, in this area at least it has not proven to be particularly fruitful for further research. Nesti (1990:425) rejects the dichotomy "visible-invisible" along with all the other dichotomies with which religion has been yoked and Luckmann is not cited very frequently by students of implicit religion.

Several themes emerge as the locus of implicit religion in these studies. One is a focus on the *lifeworld*, as described by Schutz (1973), rather than on beliefs, doctrines, or institutions. Implicit religion is found in how people live and, especially, in how they create meaning and community. Second is the role that symbols, especially metaphors, play in the search for meaning. The lifeworld is a symbolic reality (Bellah, 1970). What we call "reality" is constituted through, and mediated by, symbols which create structures of meaning.[2] Many of these structures, which form the building blocks of our culture, are implicitly religious. Third is a focus on questions of identity and authority. Like all religions, implicit religion tells us who we are, and also prescribes how we are to treat our kin, our community, and strangers. Together these themes describe phenomena which are religious, albeit not in the conventional, institutional sense.

What is Constructivism?

Constructivism is a research program developed within science, technology, and society studies. Weibe Bijker, one of its leading practitioners, defined its aims as "an integration of case studies, theoretical generalizations, and political analyses...both *to understand* the relations between technology and society and *to act* on issues of sociotechnical change" (1995: 6).

Science, technology, and society studies is itself an interdisciplinary approach. While it has a number of predecessors, such as the criticism of science made by the Frankfurt School (e.g., Horkheimer, 1947/1974, 1972) and the long-neglected work of Ludwik Fleck (1935/1979), STS became an intellectual and social movement in the 1960s (cf. Edge, 1995; Restivo, 1995). Inspired by the revolutionary work of Thomas Kuhn (1970) and Berger's and Luckmann's *The Social Construction of Reality* (1966), STS was active within both the university and the nuclear disarmament, anti-war, and environmental movements. As a research program within STS, constructivism took form in the 1970s and 1980s, inspired in part by the "strong program" in the sociology of scientific knowledge (Bloor, 1976) and a number of exemplary studies of scientific work (Latour and Woolgar, 1979; Knorr-Cetina, 1981).

Constructivism has several distinctive characteristics. First, constuctivists believe that both technology and society are socially constructed. They reject the notion that technology determines society or that society determines technology. Since, as we will see, technological determinism is the dominant way of talking about technology today, much of their work is sharply critical of this orthodoxy. Constuctivists see both

---

[2] See chapter five for a more complete discussion of how symbols create meaning.

technology and society as inextricably linked (they often use metaphors such as "seamless web" or "network" to describe their relationship). They do not speak of *the* social, *the* technical, *the* scientific, or *the* political (Bijker, 1995:13) but instead see all as interwoven constructions of social actors. None have priority over the others. The task of the analyst is to discover their interrelationships.

Second, constructivists believe that whether or not a technology "works" is what needs to be explained and cannot serve as the explanation. Both "success" and "failure" need to be treated the same, that is, "the useful functioning of a machine is the result of sociotechnical development, not its cause" (Bijker, 1995: 13). In other words, the analyst cannot assume that either nature or technical feasibility are causes or explanations, nor that there is "one right way" for a technology to develop.

While constructivism is a bold and fruitful approach to the study of technology and society, it does have some limitations. So far, its practitioners have focused most of their research on the development of technology, but have spent relatively little time on its application and use. They do not pay sufficient attention to the questions raised by feminism, and would benefit from greater integration with feminist analysis. While constructivism began as a movement for reform, (Bijker, 1995) many (but by no means all) of its practitioners have been diverted into a relativism which eviscerates social ethics and political action. And most seriously, constructivism has yet to develop an adequate theory of community. Scientists and engineers are too often seen as self-interested individuals building networks over each other (cf. Latour, 1987, 1996), which is at best an incomplete picture. As a consequence, constructivists underemphasize both the institutional aspects of technology as a practice—including institutionalized norms and values—and the formative role played by language. One of the aims of this book is to address some of these deficiencies in order to make constructivism a more useful tool for social ethics.

The Plan of This Book

This book is an analysis of our discourse, our public talk, about technology. As we debate the meaning of the new technology, we define and shape it. Some people may feel threatened by the new technology, others will try to appropriate it for themselves. I will examine the debates—and conflicts—over the meaning and appropriation of computers.

I will proceed in most chapters through a series of case studies of how people speak and write about technology, which should reveal the often implicit meanings surrounding it. My approach will follow what Paul Ricoeur (1970) calls a "double hermeneutic." Part I, "A Critique of

Technological Mysticism," is an exercise in suspicion. It criticizes what I call technological mysticism, the implicitly religious understanding of technology that has developed in our society. I will try to strip away the masks and make explicit what is hidden. Many of the symbols and myths surrounding technology are destructive and lead to injustice. Until we can liberate ourselves from their power they will frustrate our ethics and distort our policies.

Also woven through the first part is a second story, an account of the dynamics of technological change, which provides it with a systematic framework. Together with each case study of the implicit religion of technology is a discussion of some of the central ideas of constructivism which aim to place the details of the case study in a broader theoretical context.

The first chapter is an overview of technological mysticism and a clarification of the issues. It begins with a discussion of how we talk about technology, laying out several commonly encountered positions and showing how this discourse shapes the possibilities for ethical action. Following a definition of technology as a practice involving technical, organizational, and cultural aspects, I look at manifestations of technological mysticism encountered in each aspect. The mystification of the technical aspect is the "technical fix," a belief that limits all options to the technical. Organizations often conceal their exercise of power behind a "technological veil," while cultural myths surrounding technology create meanings and implicit assumptions which form the framework of our discourse.

In the second chapter, I look at how technological mysticism frames debates over technological change, using contemporary utopias as an example. Some have claimed we are going through a transformation comparable to the industrial revolution which will usher in a qualitatively different kind of society based on information technology. Their claims contain assumptions which set out options and imperatives which, once accepted, control discourse. I will critically analyze these claims, finding in their work a continuation of a form of theological prophecy.

Women have often felt inhibited by, or excluded from participation in, computer technology. The third chapter explores some of the reasons why computers are perceived as a "masculine machine." Feminists debate whether computers are essentially masculine, that is, the machines are an inherent expression of male values, or the technology has been appropriated by men through various socialization and gatekeeping techniques. The first approach sees the problem in the culture of technology practice, the second in the power relationships of the organizational aspects. Two case studies illustrate both sides of this debate.

Chapter four analyzes the language used to describe computers in the mass media. The aim is to understand some of the strategies used to stabilize and close debate, turning technology into a *black box* and making what was once problematic appear "natural" and "universal." I use a historical case study here, following ten years of reporting on computers in *Time* and other news magazines, in which 36 percent of the articles used language which was explicitly magical or religious.

The fifth chapter, "Faust's Bargain," looks at the Faust story as one of the central myths of our time. At the beginning of the scientific revolution, Francis Bacon equated knowledge with power and defined technology as a means of control. Power and control are still the dominant themes in technological discourse today. But are power and control illusions? When we look at people's experience with computers, frustration is often more frequent than the promised empowerment. We are caught in Faust's bargain and need to find the wisdom to escape.

The suspicion characteristic of Part I is only the first step, however. The recovery of meaning requires two steps. First, we have to get to the root of the problem, separating real issues from the muddle of "techno-babble" and mystification. Part I addresses the first task. Second, we have to discover theological and ethical alternatives. Borrowing a phrase from Ursula Franklin (1990), I call Part II "Redemptive Technology." This section investigates the place of values and moral decision making in the development of technology in order to lay the groundwork for an ethically responsible alternative. It tries to get behind technological mysticism to fundamental questions of meaning. The theme is that by reopening discourse we may begin to turn the direction of technological change in a more just, equitable, and democratic direction.

The sixth chapter engages two philosophers, George Grant and Frederick Ferré, and a metallurgist, Ursula Franklin, in a debate over technology and modernity. For nearly two centuries analysts have found technology at the heart of the modern world. Both philosophers agree. Grant's deeply pessimistic view, grounded in the tradition of Greek philosophy and Biblical religion, measures what we have lost. Ferré looks for ways a postmodern future may be different. Franklin offers a vision of an alternative way to practice technology. From their ideas emerge six principles for a redemptive technology.

The final chapter looks at the history of technology assessment and why it failed to live up to its promise. If technological change is to become more equitable and democratic, we will have to change the model of assessment to incorporate human agency, and therefore values, into the process. The chapter concludes by examining the place of information technology in a vision of a good society. We need to articulate such a vision if we are to make informed, ethical decisions about technology. We also need institutional support. Following Robert Bellah and his associates

(1991), we will look at institutions as the place where both technological and ethical decision making occurs.

Part I

A CRITIQUE OF TECHNOLOGICAL MYSTICISM

Chapter One

Technological Mysticism

> In the intellectual life of a society in which the mechanism of traditional faith has become corroded, new myths proliferate with the greatest ease, even though they may originate in technical advancements or scientific discoveries.
>
> Leszek Kolakowski
> *Toward a Marxist Humanism*

These are trying times for the One True Faith. Recent years have seen the destruction of some of its holiest shrines and the death of devotees and acolytes. Here and there infidels have raised their heads and unbelievers have achieved notoriety. But in spite of these troubles, throughout both the industrialized and developing worlds, the One True Faith remains ascendant. Neither Christianity nor Islam, Liberalism nor Marxism, the One True Faith is technological mysticism: faith in the universal efficacy of technology. It is a system of beliefs uniting communists and capitalists, tycoons and unionists, the rich and those who would be rich. At the moment its most potent icon is the computer.

Hyperbole? Perhaps, but the parallels between discussions of technology today and theological language are striking. The Polish philosopher Leszek Kolakowski once observed that: "Twentieth-century thinkers have done everything to keep alive in our minds the main questions that have troubled theologians over the years, though we phrase the questions somewhat differently" (1968: 9). Questions such as "Who are we?" and "What must we do to be saved?" endure, even if traditional religious answers are no longer in favour. Hidden within the language we use about technology is an implicit religion. To reveal the beliefs of this religion we will have to unlock the assumptions, metaphors, and myths of that language. This chapter will begin the process by presenting an overview of technological mysticism. First we will see how religion is implicit in the way in which technology is discussed in today's society and then investigate several case studies and illustrations of the assumptions and myths underlying this discourse.

Talking about Technology

Human beings spend a lot of time talking about technology. Gurus of the latest trends in technology are rarely absent from the bestseller lists and lecture circuits. Over the years, hundreds of authors have published thousands of books and articles about this or that technology. All this chatter should not be surprising. In a profound way, human beings are technological animals.

Stop for a minute and think what your life would be like without *any* technology. As animals, most of us would compare rather unfavourably to our house pets. Without fur we can neither stay warm nor protect our skin from the sun very well. Our teeth and nails are all but useless as weapons. Even Donovan Bailey is unlikely to outrun the family dog, nobody has senses as acute as the cat. We depend upon technology every day for our survival. Without technology we would lack food and shelter. Through our technology we are clothed and protected and employed. Indeed, as Weibe Bijker once remarked (Bijker and Law, 1992: 290), just possibly the only time people don't use technology is on a nude beach. From our earliest ancestors on the African savannah, our species has shaped and been shaped by technology.

Since technology is part of our everyday experience, we spend a lot of time and effort trying to make that experience meaningful. Over the past two centuries in the West several ways of talking about technology have come to dominate our thinking. Two of these arise out of the Enlightenment tradition, with its emphasis upon reason, science, and progress. Another characterized the Romantic reaction to the Enlightenment and puts its emphasis upon emotion, community, and nature. However, all three of these ways of talking about technology use metaphors which imply its autonomy, that is, technology is portrayed as "objective" or "independent" of human beings.

For some, technology is "applied science," the results of which are "just machines." The metaphors of this first way of speaking convey the autonomy of moral neutrality, the independence of that which is a value-free instrumentality. In this manner of speaking, technology can be used for good or evil, but one cannot blame the tools for what is done with them. Inose and Pierce, for example, express this view particularly clearly: "Like fire—and like all other technologies—information technology is a tool. It is a powerful tool, the extent to which it is used for the good or for the bad of humanity is up to human beings themselves" (1984: 186).

The second expression of autonomy, "social impact," is a metaphor of collision. Technology is described as hitting or penetrating society. As usually used, this metaphor implies that technology is the active agent of change and that people passively respond. Implicit in "social impact" is technological determinism. Technology is seen as the driving force in society, or as one prominent writer put it, "technological change induces or 'motors' social change" (Mesthene, 1970: 26). Technology is in the driver's seat. Often this manner of speaking is combined with faith in progress, so that technology is portrayed as leading (or pushing) us into a better future.

Both of the first two forms of discourse are characteristic of the Enlightenment tradition. The Romantic movement appeared in the early nineteenth century as a reaction to the Enlightenment and the industrial revolution. While their metaphors emphasize the opposition of nature and the human spirit to the machine, these critics of technology are usually equally deterministic. "Technology is out of control" became a common theme. From *Frankenstein* until today, this way of talking expresses doubt, fear, and anxiety about what technology is doing to us. In many ways, this understanding of technology is the mirror image of the previous one. Whether technology is seen as "in control" or "out of control," both of these views share a belief in technological determinism. They differ in how technology is evaluated, not in their description of its autonomy.

The language of autonomy is the most common way of talking about technology today but, for reasons we will see shortly, it is neither the most accurate description nor does it provide an adequate basis for social ethics. If the language of autonomy, in any of its expressions, is inadequate, fortunately it is not the only way to talk about technology. In recent years a fourth way of understanding technology has appeared which speaks of technology as a "practice." Arnold Pacey (1983: 6) defines "technology practice" as a human activity involving technical aspects (machines and the knowledge and skills of how to make and operate them), organizational aspects (the organizational structures and economics of using them), and cultural aspects (the goals, norms, beliefs, and values surrounding the machines and their use). The technical aspect is what we usually think of as technology, including machines and tools, of course, but also the knowledge, skills, and techniques for making, using, repairing, and scrapping them. The importance of knowledge and technique cannot be overemphasized. Many important technologies either do not involve machines or tools at all, or machinery is incidental to the technique, as in a number of medical procedures. The now-famous "Heimlich manoeuver" is a good example. Organizational aspects are the institutional dimension of technology. It includes business and union activity, as well as the users and consumers of technology. In many ways, the organizational aspect can be more important than the technical. Steam engines, for instance, were in use in European coal mines for decades before they were employed in factories, but it is the rise of the factory system which marks the beginning of the industrial revolution. Scientific management (sometimes called "Taylorism" after its founder, Frederick Jackson Taylor) is a technique of management and organization which is not dependent on any specific tools or machines, yet it revolutionized industry in the first decades of this century. Cultural aspects include people's beliefs, values, and goals. Technology is enmeshed in ideologies and myths, ranging from fads and fashions up to the deepest symbols around which people structure their identities and order their societies.

All three of these aspects are interwoven, influencing and shaping each other. Technical changes bring with them assumptions of interpersonal change and changes to social organization. Cultural and organizational aspects shape the technical as much as the limits and capabilities of the machines affect the organizations and people who use them. Questions of power, and how power is organized, are thus inherent in any discussion of technology. Indeed, all three aspects are so closely bound together that the recent literature (Bijker, Hughes, and Pinch, 1987; Bijker and Law, 1992; Bijker, 1995) uses metaphors such as "seamless web" and "network" to describe their interaction. It is as a seamless web of the technical, organizational, and cultural that our discussion of technology may begin to understand the place of technology in defining our world and permit us to begin to effectively incorporate social ethics. While most of the discussion in this book focuses on cultural aspects of technology, we always have to keep in mind that all three aspects are interwoven, influencing and shaping each other in a seamless web of interactions. Computers, after all, could not become a defining metaphor in our society (a cultural aspect), without either the machines themselves (a technical aspect) or the corporations which design, build, and market them (an organizational aspect).

This discussion of technology as a practice has several implications for understanding technological mysticism. First, because it is a practice, technology is inherently a relationship or network which means that far from being autonomous, technology is one form of social action. One simply cannot understand a technology apart from the actions of people, in any of its three aspects. For instance, an analysis of the microcomputer revolution of the eighties would have to look at the technological spin-offs from the space program, the structure of corporate capitalism with its imperatives to improve productivity, and a culture which was willing to accept peering into a cathode ray tube for long periods as a meaningful activity. It also would have to include the personalities of individual entrepreneurs such as Steven Jobs, whose skills as promoters played a key role in transforming a toy for electronics hobbyists into a major industry. It is as part of a network of technical, organizational, and cultural aspects that Jobs and the others were able to sell their computers and change society.

Second, technology is never a neutral instrumentality, for two reasons. First, as Langdon Winner (1985) reminds us, technical artifacts are the embodiment of past politics.[1] The organizational aspect of technology infuses an element of power through all aspects of technology practice. Networks involve power relationships, as allies are recruited, resources appropriated, and rival networks confronted. As an artifact, a technology may incarnate those power relationships and may bear them long after the network that created it disappears. The decision to build freeways instead of mass transit, for example, is a technological decision that could shape a city for decades after the decision makers are gone. A particularly good illustration of this is given by Winner (1985: 28-29). Robert Moses, the chief architect for New York, built the Wantaugh Parkway on Long Island with bridges so low that buses could not go under them. The reason for this was to keep low-income people and racial minorities (who would use buses rather than private automobiles) from frequenting Jones Beach. Moses is dead, but his prejudices are still embodied in steel and concrete.

The second reason technology is never a neutral instrumentality is that every technology is enmeshed in a web of values. Arnold Pacey describes three sets of values involved in the practice of technology (1983:102). The first set he calls virtuosity values. These are the values of exploration and conquest, of heroism and the thrill of discovery. Held by many engineers and scientists, these are the values of the "virtuoso" who displays prowess and expertise in mastering technique. The second set Pacey calls economic values. These are the values of management, control, and production. Here are the values of bureaucrats, managers, and production engineers. The emphasis is also upon expertise, but it is economic expertise that masters people and systems. The third of Pacey's sets, user or need values, place priority on stability, maintenance, and care for nature and people. They are more oriented towards appropriate use of technology than high-tech performance or continuous production and growth. All three value sets are found in every institution, although which are predominant will vary from one empirical setting to another. In the

---

[1] This has been a point of considerable controversy (cf. Gill, 1996; Grint and Woolgar, 1995, 1996; Winner, 1993; Woolgar, 1991). I am not making an *essentialist* argument—that a machine has an inherent logic or essence—but rather insisting that all of the technical, organizational and cultural aspects of technology be fully explored. Nor am I arguing that society determines technology. The reverse of Winner's statement—politics embodies technology—is equally true. Consider, for example, the dependence of current North American politics on polling techniques, focus groups, computerized databases, direct mail, etc. (Indeed, some charge that today's politics are little more than the application of these technologies.) Although we have to remember that political action does not always lead to the intended consequence (Berg and Lie, 1995), there is a tendency on the part of some scholars to overemphasize unconstrained individual or group agency and to pay insufficient attention to history, social structure, and power relationships. What is needed is a balanced analysis which situates individual actors within their communities and institutions (cf. Bijker, 1995).

industrialized world today, virtuosity and economic values are far more powerful in most technology practice.

So technology can never be a neutral instrument because it is *always* intrinsically bound in a web of politics and values. Those who argue that it is "just a tool" can only do so by ignoring technology as a practice. But even the most willfully blinkered approach to technology as hardware cannot avoid confronting politics and values. As soon as someone says a computer is "user friendly" or that one does "power computing," for instance, she or he is making a political statement and declaring values. There is no point in the history of a technology that it is not always already enveloped in a web of values and politics.

The third implication of the seamless web metaphor is that technology has a tradition. Tradition in technology includes, but is not restricted to, the developmental history of an artifact (such as the succession of models of a car, aircraft, or computer), the ongoing corporate culture of its organization, and the assumptions, ideologies, values, and myths that underlie discourse about the artifact. One illustration comes from the early days of railroads. Steam locomotives were developed more or less independently in the United States and Britain. In each country, the several manufacturers each in their own way produced steam locomotives which looked distinctively American or British. Each country had a "tradition" which tacitly described what a locomotive should look like. In another example, women at the end of the nineteenth century used the new technology of the typewriter to take over the largely male occupation of secretary. As a consequence, the typewriter became so gender-typed as a "woman's machine" in many business circles that some corporate executives were reluctant to use computers in the early 1980s because keyboard use required typing. Through such traditions, the cultural aspect of technology ensures that both its technical and organizational aspects are infused with norms, values, and beliefs.

Furthermore, *because* technology is a practice, it is more than what we do, it has become part of what we are.[2] Today, technology is part of our identity. In the past two hundred years, it has become common to define our species as *Homo faber*, "man the tool-maker," and to identify cultures with their technology—neolithic, Bronze Age, Space Age. If we have the latest gadgets, we define ourselves as "modern" while those who lack such are "old-fashioned," "quaint," "underdeveloped" or "backwards." From the adolescent with his car to the hacker with his computer to the general with

---

[2] Grounding identity is often seen as an important aspect of religion. Indeed, Hans Mol goes so far as to define religion "as the *sacralization of identity*" (Mol, 1976:1). Edward Bailey found identity is the central concern of implicit religion, defining it as "commitments," "personal depths," "integrating foci," and "intensive concerns with extensive effects" (1983:70-71).

his[3] latest missile, machines get tied up with our personal identities as well (Turkle, 1984a, 1995). We increasingly come to identify who we are by and through the machines we use.

Perhaps because technology so closely ties together what we do with who we are, much of our discourse about it is mystification. If you want a demonstration of just how mythical technology is, spend an afternoon looking at automobile ads or popular writing about computers. The language is that of potency and mastery. Through our machines we command the transformative power of the numinous, or at least we appear to. Yet, ironically, the more technological our society has become the more "technologically illiterate" the majority of people are. In advanced industrial society our life has become increasingly segmented through the division of labour and the "separation of spheres" (MacKenzie and Wajcman, 1985:174; Staudenmaier, 1988). On the farm, work and home were combined. As turn-of-the-century defenders of the agricultural lifestyle noted, the farm family had to be "handy" (MacDougall, 1913/1973:140). They had to repair many machines and therefore had to know how they worked. Today's urban dweller uses many more—and much more complex—machines, but now most people are consumers rather than producers of technology. Few see it built, few know what makes it work, and even repair is often a matter of replacing one "black box" with another (Tenner, 1996). Technology has taken on aspects of "nature," an independent force, powerful but mysterious. Technology itself has become more important as "display" (M. Smith, 1983:179) becoming a commodity, a status symbol, an object of conspicuous consumption. Its power is ours to use but not to understand. Technology is magic.

As practice, identity, and mystification, technological mysticism lies at the heart of advanced industrial society. When we look at technology this way, we find some remarkable similarities with theological traditions. Like a religion, technological mysticism "binds together"[4] core values into a coherent, if implicit (and often unexamined) set of beliefs and rituals. But do we want to accept it as the One True Faith? If we are unwilling to bend the knee, we must, as a first step at least, "demythologize" technological mysticism. To begin doing this I will examine several case studies and illustrations found in recent discourse about technology, looking at one for each of the three aspects of technological practice.

---

[3] The use of the masculine pronoun here is deliberate. As we will see in chapter three, many technologies have become closely identified with masculinity.

[4] Religion, from the Latin *religare*, to bind together.

The Technical Obsession

The first aspect of technology is what we usually think of when we hear the word: the machines and the "software" that controls them. Belief in the universal efficacy of technology easily leads to an obsession with the technical aspect. This is faith that a "quick fix" or "magic bullet" can circumvent any difficulty by reducing it to a technical problem (Weinburg, 1966/1986: 22). Pacey defines such a "technical fix" as "an attempt to solve a problem by means of technique alone," ignoring "possible changes in practice" which could prevent the predicament in the first place (1983: 7). The technical fix is often an attractive idea, optimistically presenting us with a (relatively) simple solution to an intractable problem. Yet the confident optimism, even utopianism, so characteristic of this discourse conceals premises which, more often than not, exacerbate the situation.

North America is full of examples of the technical fix. Canadian politicians are enamoured of "megaprojects," which are portrayed by their advocates as a quick fix to provide jobs and development. In the United States, the military, in particular, has been obsessed with technology as the answer to every security problem. A case study which provides a particularly clear demonstration of the technical fix is the debate over computer literacy in the schools.

For a long time there has been widespread concern over the quality of education in North America and elsewhere. In the early 1980s, computer enthusiasts, a lot of parents, and many school administrators saw computer technology as the way to "'fix up' our schools" (National Association of Secondary School Principals [NASSP], 1984:1). If schools had computers and students became "literate" in their use, the argument went, then bored pupils would be challenged, dropouts would stay in school, the curriculum would become relevant, and all would be prepared for high-tech jobs in the future. Like every expression of faith in the technical fix, discussion of computer literacy rested upon deeply held assumptions. Examination of these assumptions will reveal some of the bases, and the fallacies, of the technical obsession.

The computer literacy campaign displayed a great deal of mystification. In spite of a massive debate, no clear consensus emerged on what the term meant. Some definitions were extremely general, such as that offered by the US Office of Technology Assessment [OTA]: "The ability of individuals to use an information system to help them at home and work" (1982: 19). Others were detailed lists of skills, knowledge, and attitudes (Charp et al., 1982: 6). David Moursund offered what may be one of the better definitions: "All students should become computer literate. This means that they should gain a substantial level of computer awareness and develop a functional level of skill in using computers as an aid to problem solving in a variety of disciplines" (1983: 38). But why should *all* students become computer literate? Underlying all these definitions is an interconnecting pyramid of assumptions, yet through most of the debate the soundness of these assumptions was unquestioned.

The first assumption underlying the concept of computer literacy was that we are witnessing the emergence of a qualitatively new type of social order—information society. It was believed that "the computer will restructure society as completely as the steam engine did in a former age—and public education will not be spared" (Lindelow, 1983: 4). Because of this, "a computer-literate populace is as necessary to an information society as raw materials and energy are to an industrial society" (Deringer and Molnar, 1982: 3). These ideas were not unique. The work of Drucker (1968), Bell (1973), Masuda (1980), and others provoked considerable discussion about "post-industrial" or "information" society. Unfortunately, most of the debate over computer literacy did not approach information society as an hypothesis to be tested, but as a proven fact. Sweeping generalizations were common, with only the most meager data (if any at all) to back them up. Many discussions were lifted wholesale from the works of popularizers such as John Naisbitt and Alvin Toffler (for example, NASSP, 1984). The interests of business were assumed to be for the good of all (Mangan, 1992; Noble, 1991). In nearly every discussion the requirements of an information society were a premise, not a conclusion.

Yet there were some peculiarities to this premise. Most of the discussion was in the future tense. Few of the participants analyzed what the current situation actually was. Instead their discourse was full of predictions of what will be, or could be, or should be (Danzinger, 1985: 7). Only a few of these predictions were based upon the projections of a technology assessment (for example OTA, 1982 or Pogrow, 1983). Most were pure speculation and, with the hindsight of a decade, most were wide of the mark.

Additionally, many in this discussion displayed a peculiar attitude. While some spoke optimistically, or with vision, or expressed the notion that "being a change agent can be fun" (Moursund, 1983: 46), for many more there was a strong undercurrent of fear. Many were afraid we "won't be ready" for an information society, that we will be left behind by new technologies, that we will be rendered obsolete and unemployable.[5] Charp et al. were typical with their worries that "the individual who does not have a grasp of these devices can be severely handicapped (1982: 1), and that "schools as we know them today may have no place in the future" (1982: 25). This combination of vision and fear of the future was perhaps best illustrated by the NASSP (1984). They went on for pages describing the

[5] Notice the rather striking analogy to the old religious anxiety "will we be saved?"

information society and all the new high-tech jobs for which students will need to be trained. They talked about the "higher level of cognitive skills" needed for "individual productivity and satisfaction" (1984: 18). (Notice they are making the assumption that job-market preparation is the chief goal of education.) But then they noted that the same projections said that there would be relatively few of these jobs. "By 1990, only 7 percent of the workforce will be employed in the very highly technical jobs; five out of six workers, however, will be in clerical and service jobs such as waiters, waitresses, salespeople, janitors, security guards, cashiers, and cooks" (1984:17). Their "vision of the future" was one of savage competition for the few places in the technological elite. The losers in this struggle would have nothing to look forward to but low-skill, low-pay, low-status, dead-end jobs. With such a vision as a basic premise of the computer literacy debate, is it any wonder it was so often underlain with fear?

I think the reason why fear undergirds so much of this discussion can be found in a second assumption: technological determinism. Technology was seen as the driving force of social change, and computers were seen as *the* decisive technology. Klassen and Anderson were typical of much of this debate: "There is little doubt, it seems, that life in the United States and the rest of the industrial world, and eventually all over the planet, will be incalculably changed by computer technology" (1982: 26). Notice the assumptions. The computer is the motor of change. "Progress" is linear, beginning in the most technologically "advanced" countries and spreading inexorably outward to the whole planet. This "progress" will "impact" every aspect of our lives (by implication, whether we want it to or not). Thus the dilemma: "The microcomputer revolution is upon us. Almost every facet of our lives is impacted by the power of computers....As educators what can we do?" (Brombacher, 1982: 5). Since the information society was seen as the inevitable consequence of the new technology, we have to "adjust," "adapt," and "get ready for it" or we will be left behind. Discussion of whether their vision of information society was *good* or not was notably absent. After all, how can one morally evaluate the inevitability of progress? On those few occasions when values were discussed, the emphasis was on "developing positive attitudes toward personal use of computers" (Klassen and Anderson, 1982: 31). In their view, the machine is dominant.

This technological determinism led directly to a third assumption underlying the concept of computer literacy: that computers will transform teaching. It is in discussing teaching and the "technologically relevant curriculum" (Pogrow, 1983) that computer literacy enthusiasts became most utopian. Lindelow is representative:

> CAI (computer-assisted instruction) will free the classroom teacher to give individualized attention to students who need it and to concentrate on the "creative" aspects of teaching. Going back to the traditional method of teaching will become inconceivable. Computers will become—or will seem to become—indispensable to a good educational program. (1983: 8)

All sorts of confident predictions are made here, but when examined they prove surprisingly vague. The "creative aspects" are not specified. This is doubly vague since the accumulated experience of the teaching profession "will become inconceivable" once the machines come into their own. It is assumed that appropriate software will be developed, and that enough machines will be available in every classroom to permit "individualized attention." So far, these predictions have not come true. In spite of massive investments, few schools can show qualitative improvements attributable to computer literacy programs (Chandler et al., 1992, Mangan, 1992; Noble, 1991).

More ominously, once a substantial investment is made in computers, pressures mount to use them, whether quality software is available or not. David Moursund recommended: "All aspects of the curriculum should be reconsidered in light of the existence and widespread availability of calculators and computers. Modifications should be made to take into consideration the capabilities and applications of these machines" (1983: 38).

Here is a pure expression of the technical obsession. In effect he is saying that using these machines is what is most important for education, not the needs of the students or the content of the class. If computers cannot perform to the standards of the traditional curriculum, lower the standards to the level of the machine. Only great faith in the efficacy of computers could lead even an enthusiast to make such a recommendation.

Why did so many of the computer literacy enthusiasts have so much faith in the machine? I think the answer lies in the fourth assumption, that data processing provides a model of thought. Computers were believed to be effective in teaching *because* they are like the human mind. The literature is replete with anthropomorphisms. Computers are said to have memories and artificial intelligence. Some claimed that "communication with computers via computer language is analogous to interpersonal communication via ordinary language" (Klassen and Anderson, 1982: 28-29). Others even spoke of "understanding the psychology of human/ computer interactions" (Watt, 1982: 61). The conclusion was that "as the eighties progress, teaching computers will start acting more and more like the human teachers they are modeled after" (Lindelow, 1983: 18). Sometimes the roles are reversed, and humans are described in machine terms (Hunter, 1982: 33). Either way, a particular philosophy was being pushed.

This philosophy is revealed most clearly in discussions of how computers aid in "problem solving":

> Problem solving is a central and unifying theme in education. Any discipline can be framed as a hierarchical set of problems to be solved. Instruction in a discipline leads to understanding the nature of the discipline's problems: the problems that have been solved, how to solve some of the problems that have been solved, what problems have not been solved, and how to formulate and attack new problems. (Moursund, 1982: 77)

In other words, education is a process of learning algorithms.[6] Once the correct algorithm has been found, the "solution" to any problem is then a matter of processing sufficient information (i.e., facts). At this point there were some disagreements in the literature. Some placed a more positivistic emphasis on "the nature of the information we are required to process in order to understand our world," seeing changes occurring through "a consolidation and quantum jump in the data base" (Seidel, 1982: 31). Others placed more emphasis upon "procedural knowledge," usually described in terms of "acquiring skills" (Papert, 1980: 135). Both, however, understood thought and learning as directly analogous in human and computer. Technology provided the model for the mind.

The problem this poses for computer literacy is that the data processing model of the mind is fundamentally flawed. The epistemology advocated by the enthusiasts for computer literacy may have been consistent with the interests of the corporations and military which sponsored much of the research into computer-based education (Noble, 1991), but their particular model of the mind contradicted much of the philosophy of education and the experience of the craft of teaching. People do not think in algorithms, nor does data determine our ideas.[7] Indeed, our perceptions and intuitions—our ideas—shape the data. Theodore Roszak argues:

> It is only when we strike a clear distinction between ideas and information that we can recognize that these are radically different levels of discourse requiring different levels of education....The ideas that govern the data are *not*

---

6 Notice the similarity between Moursund's description of "problem solving" and what Thomas Kuhn (1970) described as "normal science." Indeed, one is often struck by the similarities between these computer enthusiasts' philosophy of education and that of the philosophy of science of fifty years ago. While this is surprising, considering how much emphasis most of them place on being "up to date," it is consistent with the origins of the concept of computer literacy. See Noble (1991) for a history of the development of computer-based education and its roots in military research.

7 In epistemology and the philosophy of science this is known as the Duhem-Quine Hypothesis, which states that theory is underdetermined by fact (Harding, 1976; Bernstein, 1983; Latour, 1987).

> information; nor are they sacrosanct matters of mathematical logic. They are philosophical commitments, the outgrowth of experience, insight, metaphysical conviction, which must be assessed as wise or foolish, childlike or mature, realistic or fantastic, moral or wicked.... Weighing such matters up in the critical scales requires an education which standard computer literacy will never provide. (1986: 119-120)

As a consequence of this contradiction, computer literacy did not fulfill expectations. After consuming enormous resources, computer literacy has had, at best, a marginal effect on the schools and, as Douglas Noble argues, "at worst, the potential impact of [computer-based education] on education insomuch as it reflects a continuation of the momentum accumulated throughout its historical development, leads only to a further fragmentation, decontextualization and depersonalization of education" (1991: 189). Like so many other technical fixes, computer literacy has failed and, indeed, may have made the situation worse.

This case study shows why the supporters of a technical fix have so much faith in the machine. If, as the enthusiasts for computer literacy assumed, computer data processing is analogous to human thought, and if this technology is the motor force of social change, then of course computers will transform both the nature of teaching and the nature of society itself. These four assumptions led to the conclusion that all students should become computer literate, and that computers should become the centerpiece of the curriculum. These assumptions also reveal both the fallacy of the technical obsession and the root of its enduring power. Devotees of the technical fix believe that the technical aspect of technological practice is the *only* aspect. Technique is the solution because the question is technical. But as the premises underlying the computer literacy debate show, this is false consciousness. Belief in the technical fix itself rests upon an entire worldview. It encompasses claims about human identity—indeed about the nature of the mind itself—about the nature of society, and about the nature and direction of social change. For believers, technological mysticism is a worldview, an implicit religion, and therefore is non-falsifiable. No number of failures of technical fixes can dim the faith, for that faith is deeply grounded in our culture. It is supported by powerful organizations as well.

Organized Mystification

Part of the reason technology is so powerful lies in its organizational aspect. Technology goes beyond machines and the techniques for using them to include all the people who plan, design, build, and run them. Technology is done by big organizations. It is embedded in the institutional structure of society and includes science, government, the military, business, and labour unions, as well as educators. All display symptoms of technological mysticism.

The type of mystification most often found in large organizations is what Herbert Marcuse called the *technological veil.* The interest of the organization "disappears behind the facade of objective rationality" (Marcuse, 1964: 32). Couched in the jargon of the technical fix, a curtain of ideology masks the values—and the power—of the organization. What the organization does is portrayed as in accord with reason and empirical reality—"there is no alternative." Although a technological veil may cloak an organization in a number of ways, in today's discussions the most prominent is the mystique of the high-tech organization. From executives acclaiming the "lean and mean" corporation ready to face "global competition" to scientists pushing for multibillion-dollar particle accelerators, the high-tech organization is touted as essential for our survival. The financial disasters at IBM a few years ago are one indication of the hollowness of such rhetoric. But why should this kind of rhetoric have such appeal? A case study will reveal the roots of its power.

The *Challenger* disaster was more than a technical failure and a tragic loss of life, it symbolizes the organizational aspects of technology. NASA is the epitome of "big science," embodying all that high tech promises and is supposed to mean. However, as the *Challenger* investigations revealed (Vaughan, 1996), NASA turned out to be just another bureaucracy. For all its high-tech image, NASA was no different than the Pentagon or the Kremlin or the post office. It had become ensnared in the organizational manifestations of technological mysticism.

Like the debate over computer literacy, discourse on space displays false consciousness. William Boot, in a review of the relationship between NASA and journalists, concluded: "The shuttle seems, in retrospect, to have been one of the biggest con jobs in recent memory—a craft without a clear purpose sold by NASA on the basis of wildly optimistic cost and performance projections. The press, infatuated by man-in-space adventure, was an easy mark" (1986: 24). The technological veil is endemic to the space program. It began long before the shuttle program, and today it continues unabated.

From the beginning, the space race was less about science than about *technological display* (M. Smith, 1983). Science and technology were packaged as ideological weapons in the cold war. Stung by the first *Sputnik* and other early Soviet successes (and the Soviet's own organized mystification about what it meant), the Americans responded with a crash program to put an astronaut on the moon. Stephen Strauss summarized:

> In the United States, pure exploration was driven by what some termed "technological anti-communism." Propelled by President John Kennedy's

> promise to beat the Soviets to the moon, astronauts walked, prospected, golfed, and leapt around like balloon-suited gazelles on the lunar surface, largely to demonstrate the attractiveness of the U.S. political-technical Goliath. (1988: D1)

This pattern of discourse was set even before JFK commenced the race to the moon. In 1959, the report of James R. Killian's Science Advisory Committee laid out the justifications for the manned space program (reported in M. Smith, 1983: 193-94). They were 1) the urge to explore and discover, to go (in the committee's words) "where no one has gone before;" 2) defence objectives; 3) national prestige; and 4) scientific advancement. These justifications (including the slogan picked up by "Star Trek") were repeated again and again throughout the Mercury, Gemini, and Apollo programs. They were recycled for the space shuttle. They are still in use today in the debates over the space station and Mars exploration.

The language of the space program persists because it is grounded in deeply held sets of values. Many scientists and engineers, and most astronauts, expound virtuosity values which emphasize the heroism of exploration and discovery. These values inform most of the public justification for "man-in-space." Economic values, held partcularly by NASA managers, are also used to justify the space program, particularly job creation and the trumpeting of an estimated 30,000 spin-offs (Strauss, 1988: D4). The third of Pacey's value sets, user or need values, are notably absent from the rhetoric of NASA and its allies.

In the discourse of the space program, virtuosity and economic values overdetermine each other. Together they help erect a technological veil.[8] Through organized mystification the media and public were "brought on side." NASA jargon proliferated, spreading to popular culture (remember A-OK?) and giving the illusion of cognoscenti status to those who could translate it. A sense of participation created a common identity, specially among journalists, who commonly used the first person plural "we" when reporting (Boot, 1986: 29). Walter Cronkite's enthusiasm was widely shared. Diane Vaughan comments on the "transcendent, near-religious" reverence for technology tapped by NASA.

> Almost at its inception, NASA's space program became a cultural icon. For many citizens, it represented and glorified American enterprise, cutting-edge technology and science, pioneering adventurism, and national and

---

[8] Although the absence of substantial user or need values at times threatened to rend the curtain. Purposeless exploration for its own sake—what M. Smith (1983:197) calls the "Columbus Principle" of discovery, that any endeavor can only be justified after the fact—the lack of planning and follow-through for the post-Apollo period, and the occasionally bitter controversies between "man-in-space" advocates and space scientists at times threatened NASA's facade.

> international power. Each awesome space achievement was a celebration of technological advance that reaffirmed the spiritual and moral values that supposedly underlay it. (1996: 387)

Thus the *Challenger* disaster was more than a tragic accident—it was a catastrophe. It became the occasion not only for mourning, but for retrospection, repentance, and rededication.

While all this was going on "in front of" the technological veil, behind it lay other manifestations of technological mysticism. Some critics claim that the problem with NASA is a *cult of procurement.* Regularly found in both the private and public sectors, the cult of procurement occurs when too many jobs—and the careers of too many decision makers—are riding on any piece of technology. Decisions are made on the basis of organizational norms and values—the corporate culture—which may be regarded outside the organization as deviant or criminal. Recent, if extreme, examples are the falsification of tests and data for the now-canceled Sergeant York anti-aircraft gun and the Strategic Defense Initiative, or the decision by Ford to continue producing the Pinto even though the tendency of its gas tank to explode in rear-end collisions was known. The way to personal advancement in these organizations is as project manager for a new technical item, thus there is a built-in incentive to keep the project going, however poorly it may perform. The bigger the budget, the more power and importance the project managers have, and so there is another built-in incentive to let costs escalate. If the machine does not perform up to specifications, the tendency is to hush it up and make "modifications." Whistle-blowers are discouraged. The end result is too often a project with a life of its own, costing more and more, performing less, and getting further behind schedule, but almost impossible to stop. To these critics, this is the tragedy of *Challenger.* Wasteful and inefficient, subject to political pressures, NASA paper-shufflers overrode their own engineers and sent seven people to their deaths in a machine NASA had been warned was unsafe.

The picture which emerges from Diane Vaughan's (1996) exhaustive study of the *Challenger* launch decision, however, is a good deal more complex. The problem was not amoral middle managers who disregarded safety under political pressure or to advance their own careers, she claims, but in the structure and culture of NASA itself.

> The *Challenger* disaster was an accident, the result of a mistake. What is important to remember from this case is not that individuals in organizations make mistakes, but that mistakes themselves are socially organized and systematically produced....Its sources were neither extraordinary nor necessarily peculiar to NASA....Instead, its origins were in routine and taken-

> for-granted aspects of organizational life that created a way of seeing that was simultaneously a way of not seeing. (1996: 394)

Over the years NASA and its contractors had produced a workgroup culture based on scientific and engineering practice and lines of bureaucratic and political accountability. Key to this culture was a scientific paradigm premised on faith in redundancy as the key to reliability and safety and on an adversarial style where every position taken must be defended in public disputation. This workgroup culture functioned within NASA's own culture of production in which scarcity of resources (post-Apollo times were lean at NASA) meant that compromises between production and launch schedules, costs, and safety were normal. All of this was encoded in the rules of procedure and accountability which guided NASA practice. The result was the construction of a definition of "acceptable" risk which was unable to perceive the danger of failure of the O-ring seals on the booster rocket. This was compounded by *structural secrecy*, by which Vaughan means "the way that patterns of information, organizational structure, processes, and transactions, and the structure of regulatory relations systematically undermine the attempt to know and interpret situations in all organizations" (1996: 238). At NASA disputes were a constituent part of workgroup culture, but only the conclusions would be passed up to the next level in the hierarchy. The attempt to avoid overwhelming upper-level managers with detail meant that they were uninformed about problems with the O-rings. The result of all these organizational factors was disaster. When the question of O-ring safety was raised by several engineers at Morton Thiokol on the unusually cold night before the *Challenger* launch, they were unable to prove their case to the satisfaction of the workgroup norms, while structural secrecy hindered communication. Vaughan concludes:

> It can truly be said that the *Challenger* launch decision was a rule-based decision. But the cultural understandings, rules, procedures, and norms that always worked in the past did not work this time. It was not amorally calculating managers violating rules that were responsible for the tragedy. It was conformity. (1996: 386)

Scheming managers did not destroy the *Challenger*, the corporate culture did.

Since the disaster (and in spite of massive inquiries) there is little evidence that NASA has learned its lesson. By looking for someone to blame (and focusing on the middle managers) the official inquiries exonerated both senior NASA management and political leaders and the organization itself. Rules were changed and personnel replaced but NASA's practice of technology remained unaltered. The norms and values of the corporate culture are still in place. The same rhetoric is invoked that has been used to justify the "man-in-space" program for the past thirty-five years. Technological mysticism *is* the corporate culture.

The questions raised by NASA's practice directs us back to some of the assumptions of the computer-literacy debate. There it was assumed that a high tech society will be *qualitatively* different. The technical fix will solve problems because technology determines the nature of society and the nature and direction of social change. But if high tech is supposed to do things differently, *where* is it happening? Perhaps no organization in the world is more high-tech than NASA, or more publicly devoted to the technical fix. Yet in the end, NASA turned out to be a bureaucracy like any other. The organizational aspects of technological mystification overdetermine the technical. Behind the veil of organized mystification are the same old values and power structures.

Culture and the Myths of Technology

Technology is not only machines and organizations but a set of attitudes and approaches to the world. This is why devotees of the technical fix, who claim technology is "just a tool," so often end up appearing naive. Technology is not value-neutral because values are an essential part of technology from the beginning. They underlie assumptions about technique and shape the corporate culture of organizations.

Five myths lie at the heart of advanced industrial society. These are central, usually unquestioned, sets of beliefs that give meaning to the world. Like all myths, they both reveal and conceal. They create meaning and ground identity on the one hand, but on the other they may obscure and mystify. Together they shape the values of technology practice and form the creed of technological mysticism. I will introduce them here, and return to them again in the following chapters.

The first myth is that knowledge is power. This is the dream of mastery and control. At the very beginning of the scientific revolution in the seventeenth century, Francis Bacon set the goal of science: mastery over nature. "Knowledge is power" was his slogan and technology is the means by which knowledge becomes power. In the early nineteenth century, August Comte's slogan "*savoir pour prevoir, prevoir pour pouvoir*" (knowledge for prediction, prediction for power) set the agenda for generations of social scientists as mastery over society in the same way that natural science has supposedly given us mastery over nature. This is also the Faust myth. As recounted by Marlowe and Goethe, Dr. Faustus was a Renaissance academic who sold his soul to the Devil in exchange for unlimited knowledge and power. The condition (according to Goethe's version) was that Faust could never be satisfied. The moment he was content and ceased his striving his soul was forfeit.

Bacon's aim was noble. His goal was to free humanity from the vicissitude of fate, to emancipate humanity by means of the domination of nature (but cf. Latour, 1993). If we had enough knowledge, he reasoned, we could use that knowledge to control nature and save ourselves from all the suffering caused by disease, want, and natural disaster. The idealism of Bacon's vision has inspired countless scientists, engineers, and technicians to continue the quest. But just as Mephistopheles twisted every exercise of power by Faust, so our attempts at mastery and control keep foundering. The state of the environment is the most blatant example of what Edward Tenner (1996) calls *revenge effects*, but there are many others. DDT would save us from insect pests and the diseases they carried, but the birds died and the bugs became resistant to the poison. Automobiles would give us freedom, comfort, and speed as we traveled, only to deposit us in gridlock on polluted streets. We split the atom to give us security, only to live in fear. Computers would put the world's knowledge at our fingertips, only to smother us in an avalanche of information. We are caught in Faust's bargain.

A second myth is that technology is gendered—men and women are supposed to have a "natural affinity" for certain machines. Many men love machines and controlling them is an important part of their identity. Women have machines too: sewing machines, stoves, typewriters (the technologies associated with women's "traditional" roles) but those machines culturally assigned as power symbols (e.g., high tech) are seen as a masculine preserve.

In patriarchal culture, men are commanded to be masters and in control of any situation. In spite of the enormous effects of the women's movement, this has not changed much. Deep in their psyches, most men still feel compelled to demonstrate mastery. Now, as Turkle (1984a: 104ff.) rightly points out, mastery is a central aspect in the identity formation of all men and women. But what she calls "hard mastery"—objectifying, abstract, dominating—is "overwhelmingly male" (p. 108). This form of mastery is just what most forms of technology are all about. Both patriarchy and technology demand virtuosity—a demonstration of mastery and control. Indeed, the "virtuosity values" described by Pacey (1983: 106) are the same as those of the male sphere. Is it any wonder that high tech is seen as masculine?

The third myth, arising in part from the first two, is the cult of expertise. The "expert" is the one in control, who has mastered the technology and knows how to use its power. As our lives become increasingly ensnared in a mesh of complexity, the practical power of the expert grows. In technological mysticism, experts take on the roles of priests, prophets, and monks. Long ago, Vico (1744/1970) commented that it was in the nature of priests to maintain a secret language inaccessible to the "vulgar." The proliferation of technical jargon, accessible only to the "expert," recreates a similar phenomenon. Some jargon is, of course, inescapable technical terminology but much more of it functions as gatekeeping devices that identify the cognoscenti and exclude those who are not initiates. Much of this terminology, particularly in the computer field, is implicitly or overtly sexist, which serves as one means to keep the conclave of experts predominately male (Keller, 1985; Turkle and Papert, 1990). Priests and prophets of various religions have practiced many forms of divination over the centuries. Today's prophets and soothsayers come from the consociation of futurologists who, though they have substituted statistics for entrails, continue to perform similar functions and, if bestseller lists are any indication, continue to enjoy similar popularity. Turkle's description of the monklike subculture of computer hackers, populated by obsessive virtuosos "caught up in an intense need to master—to master perfectly—their medium" (1984a: 207), is another illustration of the cult of expertise.

The danger of the cult of expertise is tunnel vision (Pacey, 1983: 36ff.). Expertise is no guarantee of wisdom. Drawn to problems that are technically "sweet," the impulse is to search for technical solutions first. This happens repeatedly as scientists and engineers work on "the project," concerned with the immediate technical challenge but not with the implications of the whole. But what is "sweet" may not be what is good. Albert Speer described how he "exploited the phenomenon of the technician's often blind devotion to his task" to drive the Nazi armaments industry (1970: 282). The cult of expertise may be dangerous indeed.

The fourth myth is that technology is infallible. In the Providence of technological mysticism, machines do not make mistakes. When accidents happen, its theodicy assigns blame elsewhere. Nothing illustrates this better than the nuclear industry's reaction to the Chernobyl disaster (for example, Sims, 1990). Chernobyl was the accident the industry believed would never happen. The 1975 Wash-1400 study of reactor safety (the *Rasmussen Report*) had concluded "that the likelihood of an average citizen's being killed in a reactor accident is about the same as the chance of his being killed by a falling meteorite" (Lewis, 1980: 59). The *Rasmussen Report* sparked considerable controversy and the refinement of assessment techniques (Lowrance, 1985: 142-49). Nevertheless, considering Windscale, Three Mile Island, Chernobyl, Hanford, and Savannah River, all the confident "scientific predictions" about reactor safety seem no more reliable than a Ouija board's. Has this led the industry to stop and rethink what it is doing? Of course not. The quick verdict on Chernobyl was "human error" (Medvedev, 1990: 36ff; Shcherbak, 1996). We are smugly told that in the

West reactors are designed differently, therefore "it can't happen here" (Sims, 1990: 110).

It seems "human error" is turning up on a lot of epitaphs recently. It sometimes seems almost every inquiry blames people rather than the machines (cf. Collins and Pinch, 1993: 143). This faith in technology leads the investigators to overlook that *people are part of technology.* The delusions of technological mysticism habitually induce system designers to forget the users of the machines. The results are software, tools, or work areas which are anything but "user friendly." The result is also the construction of machines so complex and dangerous that there is no room for human error. But people make mistakes. In a practical sense, Murphy's Law is true: "if anything can go wrong it will." Any technical system will, eventually, fail. People cannot be alert and vigilant all the time nor react coolly and without panic to every situation. Yet nuclear technology (among others) requires exactly that. In its quest for perfection, technological mysticism has forgotten common sense.

The final myth is that technology equals progress. Progress is perhaps the most powerful myth of our time, today's version of eschatology (Kolakowski, 1968). All industrial cultures believe in it. We live with an orientation toward the future that denigrates the past, custom, and tradition. In many debates, adequate ground for dismissing an idea is that it is "old fashioned" while what is important is to be "modern" and "up to date." Notice how quickly "Luddite" comes to our lips as a term of abuse. Moreover, the single greatest indicator of progress is technology. Big corporations have been saying it for years: remember Dupont's "Better things for better living through chemistry," Panasonic's "Just slightly ahead of our time," or General Electric's "Progress is our most important product"? One cannot find more utopian speculation than among enthusiasts for this or that technology. From Star Wars generals to computer hackers, the advocates of a technical fix see their favourite machine finally ushering in the Golden Age. For example, in the vision of futurologist Yoneji Masuda:

> We are moving toward the 21st century with the very great goal of building a Computopia on earth, the historical monument of which will be only several chips one inch square in a small box. But that box will store many historical records, including the record of how four billion world citizens overcame the energy crisis and the population explosion; achieved the abolition of nuclear weapons and complete disarmament; conquered illiteracy; and created a rich symbiosis of god and man without the compulsion of power or law, but by the voluntary cooperation of the citizens to put into practice their common global aims. (1980: 156)

Through computers, the Kingdom of God will have arrived. Technological mysticism assumes the future will be determined by the machine.

Of course there is a dark side to the myth. Others—one thinks particularly of Jacques Ellul (1964; 1980), George Grant (1969; 1986), or Herbert Marcuse (1964)—are as pessimistic as the enthusiasts are optimistic. They see no escape from a society in which every aspect of life is dominated by technique. In the "brave new world" of technological progress, many question if there will be room to stay fully human. For them, the motto of the 1933 Chicago World's Fair said it all: "Science Finds—Industry Applies—Man Conforms" (in Pacey, 1983: 25).

In their eschatology, both the advocates and critics of technology are remarkably similar. Perhaps both these views are just flip sides of the same coin. Both see a future dominated by technological "progress" and determined by technology. They only differ in their evaluation of that future.

This, then, is the creed of technological mysticism. Men, displaying the mastery of their expertise, single-mindedly pursuing the challenge of "sweet" problems, seeking perfection through their machines, are leading us into a progressive future in which all aspects of life will be under our control. Working in organizations dedicated to technology, they offer a technical fix for every problem, from education to security.

We live in a technological society. The machine, the organization, the culture of technology are part of our daily lives. But do we need to worship them?

To raise this question is itself to demythologize the implicit religion of technology. Unmasked, it begins to lose its power. If technology practice is seen as "objective reality," as a factual description of "the way things are," if the future is determined by the needs of the machine, then we have no options. All we can do is accept our fate and "adapt" or "get ready for" what the machine has predestined for us. But if we see technology practice as an implicit religion, as wrapped in myth and mystification, then it becomes discourse. It becomes one way of talking about "the real world" along side of others. We become free to weigh, evaluate, and ask questions. What *are* our ultimate values? What kind of future do we want for ourselves and our children? What is the *good* society, and how can we bring it closer? Technological mysticism shortcuts this debate by making discourse into display, policy into piety. Demystified, the machines are still there but the freedom of action becomes ours.

Chapter Two

Prophets of the Third Age

It is the spirit of utopia which conquers utopia.
Paulus Tillich
Political Expectations

We live in an age obsessed with the future. From econometricians to astrologers to TV evangelists to politicians on the stump, from learned journals to supermarket tabloids, what the future may bring is the subject of speculation. In a time of rapid social change this is perhaps not too surprising. Utopias are nothing new, but today we have spawned a new discipline, futurology, and "experts" on the future, futurologists, to institutionalize utopianism. Neither are utopias based on technology new, but the rapid development of computers and related technologies has encouraged prophecies of a utopia ushered in by machinery. While politicians, journalists, educators, and bureaucrats rush to embrace the computer as the path to the New Age, perhaps we need to pause and consider just how utopian these visions of the future really are.

The "computopia" of Yoneji Masuda (1980), Feigenbaum's and McCorduck's (1983) discussion of fifth-generation computers, K. Eric Drexler's *Engines of Creation* (1986), and Nicholas Negroponte's (1995) and Frank Ogden's (1995) visions of life on the Internet—to randomly pick five recent utopians—describe not only technological and economic change but the modification of culture and the self as well. Closer examination shows these visions to be as much religious as scientific projections. Yet at the same time, their visions are insufficiently visionary, forecasting change without transformation. Because of their technological mysticism, the possible transformative power of technology is transvalued into the extension of domination and their New Age becomes the extrapolation of the logic of the present. This chapter will analyze these utopias and the manner in which they approach the future, but first we will have to examine some of the structures through which the meaning of technology is created.

Technological Frames

If, as we said in the previous chapter, we are to talk about technology as a practice, we can no longer think about it as a collection of discrete artifacts but must see it as an ensemble made up of all the relationships between its technical, organizational, and cultural aspects. This *socio-technical ensemble* (Bijker, 1993) can be spoken of in several ways—as a seamless web or as a network of human and technical resources. Either way, emphasis is placed upon people as actors, that is, as social agents who make real decisions. To talk about technology as a practice thus precludes speaking of it either as being autonomous or as a determining force. At the same time, we cannot speak of social agency as if it were merely an act of will. People, not technology, make history but, as Karl Marx once said, people don't make history any way they please. Each person is free to act, but our actions are shaped by the physical, technical, social and intellectual resources available to us and by the norms, values and language of the institutions in which we participate. If certain resources are not available to us, they cannot be included in an ensemble. Personal computers, for example, were the stuff of science fiction until the development of microprocessors (the "chip") provided a key technical resource which was previously unavailable. PCs were the stuff of electronics hobbyists until several entrepreneurs assembled the organizational resources to bring them to the mass market. Similarly, how resources are used will be shaped by the norms (or rules) and values of both the specific institutions in which we participate and by the broader culture. Every socio-technical ensemble is thus a structure built by social agents out of the resources and rules available to them (Giddens, 1984).

Some rules and resources are so basic that they are taken for granted and often used unconsciously, what Latour (1987) and others call *black boxes*. These include the technical, social, and intellectual structures of society which, while they are themselves social constructions, are simply taken as reality by most people most of the time. We cannot, as the old saying goes, expect a fish to discover water. In other words, most of the rules and resources we use have a taken-for-granted character and are therefore invisible. When we brush our teeth in the morning, for instance, we don't stop to consider the assemblage of social and technical resources (each themselves assemblages) involved in a simple daily task. The toothbrush is an ensemble of the plastics industry, the toothpaste of the chemical industry, clean water is the result of governmental water programs, and the motivation for doing it is the product of socialization into our society's rules of hygiene and etiquette, and indoctrination by the dental profession. Technology practice is therefore an ensemble of black boxes, each of which is itself assembled from other black boxes. Much of our task as analysts is to reopen these black boxes, that is, to uncover these taken-for-granted resources and the rules by which they are employed and discover how the meaning of technology is constructed.

The meaning of any technology is established by what Wiebe Bijker calls its *technological frame*, "a combination of current theories, tacit knowledge, specialized testing procedures, goals, and handling and using practice" (1987: 168). Frames are the central means by which discourse about technology is organized. They provide "the goals, the ideas, the tools needed for action. They guide thinking and interaction. A technological frame offers both the central problems and the related strategies for solving them" (Bijker, 1995: 191-92). In the rhetoric of a technological frame we find expression of the cultural aspect of technology practice.

A technological frame is thus the sum of the ways a particular social group talks about technology. It consists of the symbols and metaphors which they use to make sense of the artifacts and techniques they employ. But a frame goes beyond being a collection of symbols. Bijker adds:

> The meanings attributed to an artifact by members of a social group play a crucial role in my description of technological development. The technological frame of that social group structures this attribution of meaning by providing, as it were, a grammar for it. This grammar is used in the interactions of members of that social group, thus resulting in a shared meaning attribution. (1987: 172-73)

Frames are similar to language itself in that they consist of a vocabulary and the grammar with which to use it. They are both resources (that is, the content or vocabulary) and the rules for using those resources which together structure discourse. A frame is therefore both reflexive and interactive, patterning the relations of technology practice while creating and recreating meaning.

The utopians we are examining form one part of the technological frame of computers and their related technologies. Their predictions both bestow meaning and call for action. To understand them we will have to uncover the technological mysticism at the heart of their frame.

Utopias, Alchemy, and Fairy Tales

To say there is a utopian element in much of the current discussion of computers and related technologies is not to denigrate it. Utopias are wishes and warnings, attempts to project a world in which the problems of the writer's society are resolved (or, in a variety called *distopia*, the problems are magnified as forewarning). People have always wished for a better life and have dreamed of magical implements to bring it about. Indeed, as Ernst Bloch (1959/1986) maintains, dreams of technological utopias originated in magic—in the wondrous implements of fairy tales and the transformative quests of alchemy.

The frames of today's utopians originate deep in our folklore. "The fairy tale is not only filled with social utopia, in other words, with the utopia of the better life and justice," says Bloch (1988: 5), "but it is filled with technological utopia." The morphology of fairy tales, according to Vladimir Propp (1928/1968), includes a "magical agent" given to the hero by a "donor." Magical agents could be anything—wondrous animals are common—but very often they are pieces of technology. To use just one example, in "The Four Accomplished Brothers" recorded by the Brothers Grimm (1945) each of the brothers sets out to seek his fortune and meets a man who teaches him his trade: a thief, a stargazer, a huntsman, and a tailor. At the end of their apprenticeships, three of the four are given magical agents: a telescope that sees "all that occurs on earth and in heaven" (p. 308), a gun that never misses, and a needle which "can sew together anything, whether it is as tender as an eggshell or as hard as steel, and not a seam will be visible" (p. 309). Through the use of these implements (and their skill) the four brothers slay a dragon and rescue a princess, transforming their lives from poverty to riches and honour. In this and other such tales, the "magic of technology" is the means of wish fulfilment.

Similarly, the transformative quests of medieval alchemists also tied together magic, technology, and utopia. Alchemists tried to transmute base metal into gold. Some probably sought no more than this—their only wish fulfilment would be a full purse. For many others, however, alchemy was a spiritual quest in which they "sought a complete scheme of things in which God, the angels, man, animals, and the lifeless world all took their place, in which the origins of the world, its purpose, and end were to be clearly visible" (Taylor, 1974: 232). In their *hermetic philosophy* the spiritual, biological, and material worlds were united. To them, the chemical processes of their craft paralleled those of life and the spirit. In their compounds and operations they saw marriage and birth, death, corruption, and resurrection. "The dream of these alchemical sects," says Bloch (1959/1986: 636), "thus remained general reformation throughout, in the sense of the restoration of the original state of paradise, and above all of leading the fallen world to Christ." Their technology was a means to a greater end. The clearest example of this was the philosopher's stone, a substance of tremendous power which, supposedly, even in minute amounts could make silver and gold out of base metals and "which had unexampled powers of healing the human body and indeed of perfecting all things in their kind" (Taylor, 1974: 66). Like the tellers of fairy tales, alchemists dreamed of living "happily ever after" but they wished on a grander scale and more clearly made technology their instrument.

Just as chemistry is heir to alchemy, fairy tales and the hermetic philosophy bequeathed their spirit to technological utopias. From Francis Bacon to Edward Bellamy and H. G. Wells, technological utopias have been a distinct genre of utopian writing (Manuel, 1966; Manuel and Manuel, 1979; Negley and Patrick, 1952; Segal, 1986). What makes them distinct is their emphasis on science and technology as the key to the perfect society. They differ from science fiction in that their dreams are not just of bigger and better gadgets but of moral and social reformation. In them, Bacon's dream of mastery over nature—and therefore mastery over fate—will have been realized. Society and human life will have become perfect. Technology itself would be domesticated: clean, quiet, harmonious (Segal, 1986: 123). These visions express the hopes of all utopians—the "demands made of an unripe reality" (Kolakowski, 1968: 71), the "anticipatory consciousness" (Bloch, 1959/1986) born of wishes for happiness. Implicitly (and often enough, explicitly) they criticize the dissatisfactions and alienation of the status quo.[1] Utopian visions express ancient dreams and desires, which have never been very far away from religion. They are also inherently ambiguous, picturing a world which is static and closed in its "perfection." With the possible exception of Bacon's *New Atlantis*, technological utopias have tended to extrapolate the technology and issues of the author's present. Actual changes in society and technology soon pass them by and their visions soon cease to be visionary, giving them a quaint feeling (well captured in Corn and Horrigan's *Yesterday's Tomorrows,* 1984). In both their dreams and hopes and in their ambiguity, today's computer utopians are part of this tradition.

Five Utopias

Yoneji Masuda is the most openly utopian and spiritual of the group under study. In his view, we are entering a new era, completely different from the past. The combination of computers and communication technologies will transform production and the nature of work. Since, he argues, the economy will be based upon the production and exchange of information, we will see a basic shift in the nature of value. What he calls "information values" (1980: 29) will replace material values as the driving force in the development of society. When this happens, every aspect of society will be recast, becoming both global in orientation and scope, and decentralized in power and decision making. The gap between industrialized and developing countries will be overcome, the environment will be protected and the arms race curbed. Bureaucracy will be replaced by participatory democracy and cooperation within voluntary communities will supplant both regimentation and anomic individualism. Governed by what he calls "time value" (p. 71)

---

[1] Although Segal (1986) notes that nineteenth and twentieth century American technological utopias expressed not so much dissent as impatience. They believed in progress through technology and were eager to get on with the job.

and "the goal principle" (pp. 125-35), people will become future-oriented (what he calls "synergistic feedforward"). They will "print a design on the invisible canvas of the future, and then...actualize the design" (p. 136). Progressing rapidly through a series of stages, information society will, early in the twenty-first century, arrive at "computopia," a "society in which the cognitive creativity of individuals flourishes throughout society" (p. 147). The nature of human life itself will metamorphose. The world will witness "the rebirth of theological synergism" (p. 154), which he describes as "a rich synthesis of God and man" (p. 156). Humanity will have entered a New Age.

Perhaps because they are American, Feigenbaum and McCorduck are much more individualistic, competitive, and nationalistic than Masuda. For them, the coming fifth generation of computers will bring artificial intelligence (a term Masuda does not use). Their book is a polemic and call to action, with the utopian elements subdued. Most of it warned of the challenges the Japanese Fifth Generation Project posed to United States technological supremacy. They are, nevertheless, utopian. Feigenbaum and McCorduck call the development of artificial intelligence (AI) the "second computer revolution." It will be as monumental as the development of agriculture, the alphabet, or the printing press. Their discussions of the future *are* somewhat vague, but if so, it is because they claim that predictions are impossible: "We stand...before a singularity, an event so unprecedented that predictions are almost silly, since predictions, by their very nature, are extrapolations from things as we know them, and the singularity called reasoning machines will change things from how we know them in vastly unpredictable ways" (1983: 236).

They do, however, offer a few hints. Medicine will be transformed by "mechanical doctors," expert computer systems which "often outperform the very experts who have programmed them" (p. 88). They will not only make diagnoses at the patient's convenience, they will bring medicine to poor countries where none now exists. Books, encyclopedias, even entire libraries or the daily newspaper will be available on microchips, each tailored to fit the individual's personal preferences. Teaching will be revolutionized through the use of "intelligent tutors" and simulation games, again molded to the individual's convenience and preference. Intelligent robots will transfigure home life, as everyone will be served by willing, obedient, and tireless servants. Finally, citing Masuda's vision, they predict the coming of AI will spark a new spirituality. They conclude:

> The reasoning animal has, perhaps inevitably, fashioned the reasoning machine. With all the risks apparent in such an audacious, some would say reckless, embarkation onto sacred ground, we have gone ahead anyway, holding to what the wise in every culture at every time have taught: the

> shadows, however dark and menacing, must not deter us from reaching the light. (p. 240)

The Japanese will do it, they claim. Feigenbaum and McCorduck urge Americans to follow.

K. Eric Drexler's *Engines of Creation* is the most magical of these utopias, evoking comparison more with medieval alchemists than with Bacon or H. G. Wells. For Drexler, AI is only one part of the technological future, albeit a central one. The key to the New Age, including AI, is what he calls nanotechnology, that is, "technology based on the manipulation of individual atoms and molecules to build structures to complex atomic specifications" (1986: 288). Nanotechnology, he contends, will permit engineers to build anything, cheaply and efficiently. The structures of the brain could be duplicated, although on a much smaller—and hence faster—scale, opening the door to artificial intelligence. With AI, nanomachines (which he calls assemblers and replicators) will be able to design, build, improve, and reproduce themselves. The door to utopia will be open. "Assemblers will be able to make virtually anything from common materials without labour, replacing smoking factories with systems as clean as forests. They will transform technology and the economy at their roots, opening a new world of possibilities. They will indeed be engines of abundance" (p. 63). As an example, he describes growing a rocket engine made of synthetic diamond and sapphire in a chemical vat. Nanotechnology is the philosopher's stone.

In Drexler's utopia, humans will expand throughout the solar system and then to the stars. Cell repair machines will virtually eliminate disease and give people, if not immortality, a life span measured in centuries or millenia. Not only will the environment be protected but DNA manipulation will allow the recreation of extinct species. An information system he calls hypertext (but which looks suspiciously like Gordon R. Dickson's *Final Encyclopedia* [1984]) will place all knowledge at our fingertips. And all this will happen within the lifetimes of many people living today.

Considering that Feigenbaum and McCorduck and Drexler place such emphasis upon artificial intelligence, it is surprising that none of them define the term. To the former, "the essence of the computer revolution is that the burden of producing the future knowledge of the world will be transferred from human heads to machine artifacts" (1983: 40). What they mean by "knowledge," and how it differs from "information," is not defined either. After more or less dismissing all arguments over machine intelligence, they declare: "Intelligence was a political term, defined by whomever is in charge" (p. 45). This does not deter them, however, from concluding later that machine intelligence is "faster, deeper, better than human intelligence" (p. 236). Drexler's approach is similar. He admits that no machine can pass the "Turing test."[2] but he leaps that hurdle by simply dismissing the relevance of Turing's test and of anyone who doubts the inevitability of AI as a "biochauvinist." He concludes: "Eventually software systems will be able to create bold new designs without human help. Will most people call such systems intelligent? It doesn't really matter." (1986: 75). Such remarks indicate that these authors are not offering scientific analysis and logical projections but statements of faith.

If artificial intelligence was seen as the key to utopia in the mid-1980s, in the mid-1990s it is the Internet. Frank Ogden, who calls himself "Dr. Tomorrow" in his syndicated column, is a prophet of apocalyptic technological change. Indeed he claims that, "I'm the apostle of change. I'm the seer of change. I try to peer into the future to better understand the present" (1995: 1). For Ogden, the pace of technological change is exponential. By 2005, he predicts, 90 percent of the goods and services we use will have changed and 60 percent of today's Fortune 500 companies will have disappeared. Even more, "within ten years, technology that is hardly out of the starting gate will change 90 percent of our culture and society, as well as the way we think, learn, love, understand, work, and survive" (p. 3). *Navigating in Cyberspace* portrays itself as a guide to survival amid such change.

The key to understanding this change is the Internet. "The Net," he says, "will change society, culture, hierarchies, economics, and the way we communicate and think" (pp. 20-21). The World Wide Web will give every individual access to unlimited information. It will create a new community in cyberspace, populated by a new aristocracy he dubs the *digirati*—those with the competence, initiative and entrepreneurial skills to use the Net to survive and prosper. Everyone else will become *technopeasants*.

The digirati are destined to have a fabulous future. "Technology will make many of today's problems vanish," he proclaims, "Need I repeat all those worries from the past—nuclear conflagration, oil shortages, global warming, ozone holes, and so forth—that just disappeared?" (p. 39). Direct democracy over the Internet will end government as we know it, computers will change the meaning of intelligence and education, and jobs will become a thing of the past as people create their own lifestyles. Eventually we will even learn to control robots with our thoughts. For those with the right attitude and the wisdom to follow his preaching, Ogden promises a Golden Age.

---

[2] One measure of AI, proposed by Alan Turing in 1950, would be if a machine could carry on a protracted conversation with a person and the person could not tell if she or he was conversing with a machine. See Churchland and Churchland, 1990; Searle, 1990.

Compared to Ogden's breathless prose, Nicholas Negroponte's *Being Digital* (1995) is almost sober and conservative. In this collection of his columns from *Wired* magazine, he argues that technology is shifting from moving "atoms" to moving "bits," that is, from transporting physical objects to sending digital electronic signals. This change is "irrevocable and unstoppable" (p. 4), is increasing exponentially, and will soon transform our lives. Negroponte sees two keys to this transformation. The first are *headers*, or "bits about bits," information which tells us about other information. The second are *interface agents*, computers which will act as intermediaries with other computers. Together these will give us the ability to manage and control the masses of information made available by the Internet and to interlink not only computers and communications media but *all* our machines and appliances.

As we move beyond today's crude attempts at speech recognition to full-speech capacity and personality for computers, the machines will be freed from any limits to size or shape imposed by interface devices (e.g., keyboards, monitors, etc.). "Early in the next millennium," he says, "your right and left cuff links or earrings may communicate with each other by low-orbiting satellites and have more computer power than your present PC" (p. 6). When this happens, computers will become ubiquitous. This will transform life: "Tomorrow, people of all ages will find a more harmonious continuum in their lives, because, increasingly, the tools to work with and the toys to play with will be the same. There will be a more common palette for love and duty, for self expression and group work" (p. 221).

Work and home need not be separated, indeed, people will not be tied to any sense of place. New forms of community and government will emerge as today's children, raised in the new environment, will overcome the problems of war and want that divide us now. "Digital technology can be a natural force drawing people into greater world harmony" he concludes (p. 230).

Chiliastic Visions

Although each of these works differs considerably from the others, all five demonstrate structural similarities in their arguments. While they all talk about the changes science and technology will bring to the twenty-first century, the structure of their discourse follows a pattern of religious prophecy which appeared in the High Middle Ages. Their technological frame has as its template prophecy in the Joachite[3] tradition—a chiliastic[4] vision and call to commitment in order to usher in the New Age (Bloch, 1959/1986; Cohen, 1970; Mannheim, 1936; Manuel and Manuel, 1979). We can see this structural pattern in five ways. First, in all their visions the future is determined; second, individual identity is transformed; third, a way is offered to a "this-worldly" salvation; fourth, a call to commitment is made that is not only a break with the past but a choice between heaven or hell. Finally, all of this is presented within a context of immediacy—the New Age (usually the Third Age) is even now breaking in upon the present; the choice for salvation must be made now.

All five of these works display a faith in predestination which would shame the most unreconstructed Calvinist. Technology determines the future. Masuda does not even raise the question of *whether* there will be an information society, it is only a question of when. He argues that right now we are in the fourth and final stage of the development of computerization,[5] the information revolution is proceeding at from 3.5 to 6.4 times as fast as the industrial revolution, and that by the end of the first decade of the twenty-first century the metamorphosis will be complete. Feigenbaum and McCorduck are grandiloquent, speaking in terms of "a manifest destiny in information processing, in knowledge systems, a continent we shall all spread out upon sooner or later" (1983: 153). They invoke "the Chardin-esque sense of something one unfolds toward, not in unbending lines but languidly, pausing in the byways, the grand plan apparent only in retrospect. But inevitable for all that" (p. 58). Drexler is more insistent, and more hurried. "The technology race will continue whether we wish it to or not" (1986: 20). Attempts to curb technology are futile and counterproductive. He sees the first generation of nanotechnology by the end of the 1990s and the "breakthrough" within our lifetime. Ogden's belief is straightforward: "Technology today makes the laws and breaks the laws" (1995: 39). Negroponte is similar: "Like a force of nature, the digital age cannot be denied or stopped" (1995: 229). For both apostles of the Internet, technological change is exponential, that is, it is constantly doubling, which means that insignificant factors today may have monumental effects

---

[3] Joachim of Fiore (1145-1202), an Italian abbot, developed a new technique for interpreting scripture which found in the Bible a concealed meaning which provided the key to interpreting history and forecasting the future.

[4] Chiliasm is a form of apocalyptic prophecy which emphasizes the immediacy of Judgment Day and the coming of the Kingdom of God.

[5] The first stage he calls "Big Science-Based Computerization," the second is "Management-Based Computerization," third is "Society-Based Computerization" in which computers are "used for the benefit of society as a whole" (1980: 38). We are now entering the fourth stage, "Individual-Based Computerization."

tomorrow. Both place their predictions in the first decades of the next century. For all five of these utopias the future is set. As prophets, they are only reading the "signs of the times" for the judgment to come.

The immanent arrival of the New Age is of immediate personal concern. These utopias are not a remote or abstract ideal in the manner of More, Campanella, or Bacon but are even now breaking into the present to affect our individual lives. Present identities will be transcended and transformed as we reach towards immortality. For Masuda, *futurization* (the actualization of the future; 1980: 136) will proceed through several stages. First is individual futurization, as each individual realizes her or his own goals, which will soon be transcended by group futurization as people form voluntary groups to better actualize the future and escape the contradictions of unregulated individualism. This will evolve into community futurization, in which people will "carry on life together with a common social solidarity" (p. 139). In the end "man approaches the universal supra life, with man and god acting as one" (p. 155). Individual identity will be transcended even as it is realized.

Feigenbaum and McCorduck claim "the computer will change not only what we think, but how" (1983: 47). Many people are afraid of AI, they contend. To this group:

> Artificial intelligence threatens, in a deep and none too subtle way, their view of themselves. As humans we've taken our very identity from our intelligence. The notion that something else—what's worse, a creation of our own hands—might also be intelligent requires a considerable realignment of our self-image. (p. 44)

But for those who overcome their fears and prejudices, AI offers much. "The extension of human intellect that the Fifth Generation will give us is simply staggering" (p. 211). AI researchers, of course, are in the vanguard, experiencing the "exhilaration" and sense of self-transcedence that comes with pursuing "a goal for the good (and perhaps salvation) of their nation" (p. 152). Even more is the "immortality syndrome" (p. 86), the prospect that *their* experience will become the basis of an expert system.

Drexler does not stop with half-measures. His promise of immortality comes close to being the real thing. Through "biostasis" (cryogenics and related techniques) and cell repair machines, people—including many alive today—can prolong life indefinitely. And while he demurs that "to speak of 'immortality' when the prospect is only long life would be to ignore the facts or misuse words" (1986: 139), a lifespan measured in centuries would seem to be good enough.

Ogden's vision is even more sweeping. Computers will bring a new form of intelligence. He claims that, "Geneticists may someday discover that as the Internet world grew, and as vast Niagaras of electronic information drenched its citizens, they were genetically transformed, developing a higher level of perception and consciousness. In the future, Internet will affect the way we perceive the world" (1995: 20).

Individuals will be transformed, and so will the human species through the application of genetic engineering to eugenics. "We are at the dawn of the age that will find us creating our own successors," he claims. "Soon children will be designed who are far healthier than their parents ever were, and they will be scientifically conceived, genetically altered, disease-resistant, and specially nurtured for delivery from artificial wombs" (pp. 137-38). Within the space of a single lifetime a new human species will be evolved.

Negroponte sees the Internet "creating a totally new, global social fabric" (1995: 183) and radically transforming everyday life. With it will come the apotheosis of individualism. Today people are not treated as individuals, even with narrowcasting. We are reduced to a list of categories at which broadcasters aim their marketing spiel. In the future, digital technology will enable each person to be a market of one, selecting, choosing, and shaping the communications we wish to receive. Technology will truly empower every individual, creating the basis for entirely new concepts of community.

All that is necessary to participate in these transformations is to have faith in the vision. The machine offers the Way to God, or if not God, at least to salvation. Salvation is a common word for these utopians. What is offered is a way out of today's problems: pollution, overpopulation, poverty, war, alienation, ignorance, and every form of want. The future will be a paradise on earth. But the other side of salvation is also present: unless we seize the proffered Way we will be judged and damned. Unless we grasp the opportunity to use the technology to build utopia, that same technology will create hell on earth. To Masuda, "we cannot escape the need to choose, before it is established, either 'Computopia' or an 'Automated State'" (1980: 152). If the latter is allowed to occur, "all citizens could be completely regimented and deprived of their freedoms" and "all citizens would suffer decisive psychological, social or economic and political damage" (pp. 114-15).

Feigenbaum and McCorduck are somewhat less apocalyptic, but the choices are equally clear. The main threat is to the nation. "In one of the most important competitions the United States has ever engaged in, it seems to be losing" (1983: 186). Unless the US immediately follows the authors' program, they argue, it will go the way of Britain and be supplanted on the world stage by Japan. The threat is also to those individuals who do not follow the Way.

> In short, no plausible claim to intellectuality can possibly be made in the near future without an intimate dependence upon this new instrument. Those intellectuals who persist in their indifference, not to say snobbery, will find themselves stranded in a quaint museum of the intellect, forced to live petulantly, and rather irrelevantly, on the charity of those who understand the real dimensions of the revolution and can deal with the new world it will bring about. (p. 212)

Drexler, as usual, doesn't mince words. With nanotechnology "we will be able to remake our world or destroy it" (1986: 14). He goes on:

> Advanced AI systems could manoeuvre people out of existence, or they could help build a new and better world. Aggressors could use them for conquest, or foresighted defenders could use them to stabilize peace. They could even help us control AI itself. The hand that rocks the AI cradle may well rule the world. (p. 76)

It is essential for our salvation, he argues, that the democratic nations become the "leading force" in the new technologies.

Ogden promises personal salvation in the technologically determined future. In his vision people will be divided into two classes; "the Knows" and the "Know-Nots," the digirati and the technopeasants. He even offers a chart listing seventeen ways in which the lives of the digirati will be superior to the others (1995: 89). The question his book answers is how to become one of the Elect. The principle Sign of Grace Revealed is attitude: "Although today's rush into tomorrow is being dictated by technology, attitude is a factor that is not being given equal time in the current equation for success" (p. 17). Only optimistic, idealistic, maverick entrepreneurs will enter the Kingdom.

Negroponte's work does not quite follow the pattern here. Technology will certainly bring "this-worldly" salvation, but for him technology is such a determining force that there is little need to call for conversion and less danger of a technological hell. Perhaps this is because the people who will inherit the New Age are today's children. The fundamental division in society is generational, he claims, "the haves and have-nots are now the young and the old" (p. 204). As our children grow up with the Internet, the transformation will occur. Nevertheless, his work is still infused with the spirit of chiliasm.

> Being digital is different. We are not waiting on any invention. It is here. It is now. It is almost genetic in its nature, in that each generation will become more digital than the preceding one. The control bits of that digital future are more than ever before in the hands of the young. Nothing could make me happier. (p. 231)

Our children *will* live in the New Age, even if some of the older generation might be left behind.

The stark choice between a "this-worldly" heaven or hell in four of these works gives them a sense of *Kairos* (Tillich, 1971)—*now* is the moment of decision. Each of them is a call to commitment. The world hangs in the balance and each person must break with the past and choose the road to salvation. Masuda, Drexler, and Ogden even include the addresses of their various organizations so that the reader may enlist at once. While Negroponte's vision is not quite so stark, it too is infused with *Kairos*. The future is *right now* breaking into the present and the transformation is at hand.

There is one final continuity with the chiliastic tradition. Joachim of Fiore saw the millenium as the "Third Kingdom" or "Age of the Spirit," following upon the "Age of the Father" and that of the Son. Cohen (1970) argues that this symbol has become a staple of chiliastic movements, from the Third Rome[6] to the Third Reich. Four of the five utopias we are examining see the future as the "Third Age" (Drexler is not clear on this point). To Masuda and Feigenbaum and McCorduck information society is the "Third Age," following upon agricultural society and industrial society. Ogden is not as historically precise, offering only a generalized "The Way Things Were," "The Way They Are," and "The Way They'll Be" (1995: 34). Negroponte offers a variation. For him, we are currently living in the information age, the past is the industrial age and the digital future is the post-information age. For all five utopians, their visions are the culmination, and climax, of human history.

Visions without Vision

There is a great deal one could criticize in evaluating these positions. Indeed, while sharp attacks have already been made on both scientific and philosophical grounds (see, for example, Dublin, 1989; Roszak, 1986; Stoll, 1995; Weizenbaum, 1985)[7], I want to restrict my analysis to the contradictions of their technological mysticism. While at first glance these utopians may seem farfetched, fanciful, even magical, they seem less so when one probes beneath the surface. Their visions are insufficiently visionary. True, they do anticipate overcoming many of today's problems. In a world of war and hunger, they offer peace and plenty. But the *manner in*

---

[6] Moscow. In Russian Orthodox millenarianism, Constantinople was the Second Rome, Moscow was the third.

[7] Although *Scientific American* initially reviewed Drexler favourably (Dewdney, 1988; "Nanofuture," 1990) more recently it has been sharply critical, refering to Drexler's movement as "a latter day cargo cult" (Stix, 1996: 99).

which they foresee these changes occuring is an extrapolation of the logic of the present. In fairy tales and alchemy, and even in somewhat more conventional utopias such as Bacon's, technology has a radically liberating effect. It ruptures the present and overthrows the power of domination. In a technological civilization like our own, however, this transformative power of technology is lost. What is offered here is not the *Novum* (Bloch, 1959/1986), something radically new breaking into history, but the triumph of the same kind of technological possibilities that define the present. For all the chiliastic form their prophecies take, the substance of their vision is built upon the deep structure of advanced industrial society.

Masuda has the clearest vision of social change in his computopia and makes the most sweeping claims for spiritual renewal. Yet in its basic principles, computopia looks more like the realization of the present than something fundamentally new. Stripped of computer jargon, what are "synergistic feedforward," "time-value," and "futurization" but old-fashioned bourgeois planning ahead? Calls for individual self-realization, freedom of decision, and equality of opportunity (1980: 148-49) have been the standard fare of liberalism for several hundred years, while his talk of voluntary associations and solidarity echoes Durkheim. Even his call to live in harmony with nature betrays a deeper commitment to its domination. We will work together with nature, he says, but "not by a spirit of resignation that says man can only live within the framework of natural systems" (p. 155). But this is nature without autonomy, nature in harmony with humanity because nature is fully under control. The image of nature here is not wilderness but a well-kept bonsai garden.

Feigenbaum and McCorduck and Drexler are even more under the sway of the logic of the present as they advocate nationalism, capitalism, and the military. It is significant that in a whole book dedicated to a nationalist argument not to let the Japanese get ahead, Feigenbaum and McCorduck claim the only group in the US "which deserves credit for its enlightened scientific leadership" (1983: 215) is the Defence Department. Drexler's view is similar, up to and including an endorsement of the concept of the Strategic Defence Initiative (1986: 198-99).

The apostles of the Internet are no different. Much of Ogden's vision seems to be borrowed from the agenda of the New Right. Entrepreneurs are his heros, while the villians of his work are government, universities, and especially labour unions. The one government which receives his unqualified praise is the puritanical dictatorship of Singapore. While he casually mentions the erosion of traditional values (usually described as obstacles to progress), his prime criteria for evaluation is financial success. And given the experience of eugenics in the 1940s, his prediction that through genetics "there will be new scientific, and harmless, means to deliver physically improved future citizens into a progressive society" (1995: 137) is frightening.

Negroponte's vision seems to be the wishes of a wealthy man who wants to be left alone. For him, much of the attraction of the new technology is that it will eliminate the need for going into the office or "engaging somebody needlessly in on-line discussion" (1995: 187). Repeatedly he returns to the theme that interface agents will be able to screen and select communications so we only need see and hear who or what we want.[8] True, he does say that our children will find new forms of neighborhoods and communities, but these references are brief and abstract. His ideal embodies what Robert Bellah et al. (1985) has called a *lifestyle enclave*, a coterie of like-minded (and socio-economically similar) individuals who only associate with others like themselves. Extreme individualism is nothing new in liberal philosophy, but Negroponte seems to idealize the condition which Emile Durkheim called anomie, a pathological state of normlessness and dissociation from society.

In all five of these works, then, the deep structure of technological civilization underlies the utopian vision. These authors have faith in technology and prophesy its possibilities with chiliastic fervour. But their technological mysticism is, in Mannheim's terms (1936), more an ideology than a utopia. They do not in fact rupture the present but rather extend the current structures of domination. For all their superficial projection of renewal, what they envision under the surface is more of the same.

Overcoming Utopia

These utopias were published over a fifteen-year period, from 1980 to 1995. The earliest three—Masuda, Feigenbaum and McCorduck, and Drexler—are already suffering the fate of most utopians as their work ages and actual developments pass them by. The future, which they saw as breaking into the present at that very moment, has receded. The information society remains elusive, at least for those who expected something qualitatively different, while artificial intelligence and nanotechnology have never escaped from the realms of laboratory and science fiction. Today it is the Internet which is being hyped as the world-changing technology, and when the excitement passes (critical comment is already appearing, cf. Eisenberg, 1997; Gibbs, 1997; Holmes. 1997; Stoll, 1995; Technorealism, 1998) it will not be long before something else takes it place.

Yet these utopians are emblematic of persistent themes in technological discourse today: faith that technology is both the question and the

---

[8] He doesn't address the question of whether or not a person who preselects the information thus received to conform to a predetermined point of view is actually "informed."

answer, and belief that our machines can bring a perfect world within reach. These themes persist because they are built upon the myth of progress and a cultural legacy descended from fairy tales and alchemy. Futhermore, in many expressions, they are structured by a centuries-old tradition of chiliastic prophecy. Their persistence helps to set the agenda for discourse by acting as a template upon which new technological frames are constructed, thereby becoming a pervasive (and, ironically, traditional) way to understand invention. In other words, these themes provide the "grammar" which orders the "vocabulary" of technical innovation, structuring the potentially disruptive experience of new technology into familiar (if implicit) patterns.

The devotees of technological mysticism tout each innovation as the incarnation of progress which will usher in the New Age, conveniently forgetting that all these claims have been made before. Perhaps the best response to these visions is to recall Tillich's words: "It is the spirit of utopia that conquers utopia" (1971:180). We need to break the bonds of the present without getting caught in a static model or extrapolation of the present. We need to construct a more just and equitable future, not just fantasize about one—while remembering that fantasy plays an important role *in* the construction of such a future. I will offer my own suggestions about such an alternative in the second part of this book.

Chapter Three

The Masculine Machine

> Computing is more than a set of skills. It is embedded in a social system consisting of shared values and norms, a special vocabulary and humor, status and prestige ordering, and differentiation of members from non-members. In short, it is a culture.
>
> Kiesler, Sproull and Eccles
> "Pool Halls, Chips, and War Games"

Technological mysticism, like all religions, functions to ground identity. As we saw in the first chapter, we are technology-using animals and much of our personal identity is wrapped up in the machines and techniques which we use. This produces what may at first appear to be an ironic aspect of our discourse—many of our technologies are gender-typed. While both the Enlightenment and Romantic traditions talk about technology as if it were separate from (and for the Romantics, hostile to) human personality, in fact many machines are perceived to have sexual associations. Automobiles, firearms, and aircraft, for example, are identified with men, typewriters, sewing machines, and cooking utensils fall within women's sphere. Computers are seen as a masculine machine. Why are some technologies associated with gender, and how do machines acquire these types?

The reasons why computers are identified with men are not obvious. From Ada Lovelace to Grace Hopper and Adele Goldstine, women played a pioneering role in the development of computer science (Perry and Greber, 1990). Women do not differ from men in their skills and aptitudes for computing (Anderson, 1987; Clarke, 1986; Clarke and Chambers, 1989; Munger and Loyd, 1989; Turner, 1984). Millions of women use computers in their work every day. As a *cultural artifact,* however, computers are masculine machines. The computing culture is decisively shaped by white, middle-class, able-bodied males. Men have generally been successful in imposing their definition of what the technology means and how it should be practised, that is, they have shaped the technological frame. This can be demonstrated in several ways.

In the labour force, women frequently use computers but rarely define the computer culture. As secretaries, clerks, and in data entry, women make up a significant percentage of computer users but computerization has resulted in both a major reduction in the number of these jobs and serious deskilling of those that remain (Menzies, 1996). Women are seriously underrepresented among programmers, system designers, and others who shape computing practice. According to the National Research Council (1994), women comprise only 12 percent of the overall scientific and engineering labour force in the United States. Only 4 percent of computer scientists are women (Advocates for Women in Science, Egnineering and Mathematics [AWSEM], 1997). Only about one in four computer operations managers are women (Adelson, 1997).

Trends in higher education indicate that these labour force patterns are not likely to change soon. Although both government and industry offer incentives to encourage women to study computing (Chandler et al., 1992: 67-68), these programs have met with only limited success (National Research Council, 1994; Grossman, 1998). For example, according to the National Center for Education Statistics (1997) in the United States only 1.2 percent of entering university women planned on studying computer science in the fall of 1996. In Canada, in 1995 only 0.8 percent of female undergraduates were computer science majors (Statistics Canada, 1996). Of those students who do study computer science, slightly over 80 percent are male, a trend which has held for over a decade. In fact, the number of female undergraduate computer science majors in Canada *fell* from 1981 to 1988 and has hovered between 19 and 20 percent ever since (Statistics Canada, 1986, 1992, 1996).

Women do not choose to study computing because many perceive the computer culture as alien, if not hostile. The prevalence of white males' norms and values is well documented in the literature on socialization into computing. Children's earliest introduction to computers often begins teaching them stereotypes (Chen, 1987; Yelland, 1995). Over the past decade and a half, an enormous international literature has described the way that teaching computers in the schools discourages young women: from teachers' and students' attitudes, to teaching styles, to boy-dominated after-school computer clubs, to computer studies textbooks, to career guidance (for a few examples, see Barron, 1996; Becker and Sterling, 1987; Culley, 1990; Etzkowitz et al., 1994; Gunn, 1994; Hawkins, 1985; Jacobs, 1988; Jennings, 1986; Lee, 1995; Lewis, 1985; Makrakis and Sawada, 1996; Marshall, Erickson, and Vonk, 1986). Over the same period, the content of software has also been a major concern (for example: Frenkel, 1990; Kiesler, Sproull and Eccles, 1985; Linn, 1985; Pearl et al., 1990; Rosenthal and Demetrulias, 1988; Spender, 1995; Wise, 1997; Yeloushan, 1989). Most children's first experience with computing is with games, a market which is 80 percent male (Adelson, 1996). Computer games are usually violent and frequently sexist, reinforcing attitudes that women do not belong (Funk and Buchman, 1996; Littleton, 1993; Miller, 1996; Valenza, 1997). Games aimed at girls often reinforce stereotypical roles (Adelson, 1996). Even educational software frequently displays gender bias (Biraimah, 1993; Chappell, 1996; Hodes, 1996; Mangione, 1995). Nor does the Internet appear to be significantly different. About one-third of Internet users are estimated to be women (Adelson, 1997; Computer Professionals for Social Responsibility, 1995), but gender stereotypes and on-line harrassment are major concerns (Mahoney and Knupfer, 1997; Spender, 1995).

The fact that the computer culture is dominated by male norms and values is well documented. Even though millions of women use computers, computing practices continue to be defined and shaped by men. The key question is why?

There are two major lines of argument over why artifacts have gender. The first contends that computers are essentially masculine, that is, the practice of computing has been decisively shaped by male values and that women's values have been excluded. The second believes the problem is that computers are not so much essentially masculine as that they have been appropriated by men. By this they mean that women have been excluded from computing through a variety of socialization techniques and gatekeeping devices. Their exclusion is an act of patriarchal power both at the level of socialization and the level of discourse. The first argument sees the cultural aspects of technology practice overdetemining the technical, the second understands the problem as arising primarily in the organizational aspect. I will use case studies to examine both these arguments. I will conclude the chapter with some further reflections on technology and gender.

Is Technology Essentially Masculine?

One explanation for why computers are identified with men contends that men and women have different sets of values and cognitive styles.[1] According to this argument, the dominant construction of science and technology is patriarchal in its very essence. Computers are essentially masculine because the practice of computing has been decisively shaped by male values, language and modes of thinking. Women's language and values have been excluded.

---

[1] Not all these arguments are feminist. Some (e.g., Dambrot et al., 1985) argue that women lack the mathematical abilities to be successful in computing. However, I will only examine feminist versions of the essentialist argument here.

The roots of this, some feminists contend,[2] lie deep in Western culture—especially in Christian theology.[3] In *The Politics*, Aristotle argued that only free men fully possessed the faculty of reason. In slaves and women reason was absent or inoperative. Later Greco-Roman culture added sharp dichotomies between the body and soul and between spirit and matter (nature). Men defined themselves as spirit and soul and identified women with the body and nature. The early church appropriated all these ideas into its theology. Theologian Rosemary Ruether summarizes:

> Classical Christian spirituality viewed man as a "rational spirit." The male alone was said to fully possess this "human nature" in its essence. The male alone was made in the "image of God," modeled in his inward being after the intellectual Logos or "mind" of God (which was also the theological identity of Christ). The female was said to lack this full "image of God" in herself, and to possess it only when taken together with the male "who is her head." ...Male-female dualism was seen as a social extension of subject-object dualism, so the male alone was the perceiver and the articulator of the relationship, while the woman was translated into an "object" in relation to this male perception and "use." Thus women were seen, literally, as "sexual objects," either to be used instrumentally, as a "baby-making body," or else to be shunned as the incarnation of tempting, debasing "sensuality." (1972: 19)

For early church theologians, and especially for Augustine, the pleasures of the flesh—most fundamentally the arousal of sex—were the essence of sin because they represented the surrendering of the spirit to nature. From "the perspective of the male eye," women literally "incarnated" a threat to the redemption of a man's soul and therefore came to symbolize sin (Ruether, 1972: 100). The solution recommended by the early church was asceticism. In essence, asceticism was an attempt to use reason to maintain control: to master both one's own body and all those temptations (especially women) which threatened the loss of control.

These theological ideals became the institutionalized practices of the medieval church (Noble, 1992). In the early modern period Protestantism took asceticism out of the monasteries and into everyday life (Weber, 1905/1958). Ascetic patriarchy also dominated the universities,

---

[2] Feminism is a large and complex social and intellectual movement (cf. Mitchell and Oakley, 1986; Tong, 1989). Only a few major arguments can be addressed here. For more complete discussions of women in science and technology, see Alcoff and Potter, 1993; Haraway, 1991; Harding, 1986, 1991; Keller, 1985, 1992; Kramarae, 1988; Schiebinger, 1989.

[3] Other societies had, of course, their own approaches to technology (cf. Pacey, 1990) and their own ways of subjugating women. The limitations of time and space will restrict me to a discussion of Western culture.

which grew under clerical sponsorship. David Noble comments that "Western science thus first took root in an exclusively male—and celibate, homosocial, and misogynous—culture, all the more so because a great many of its early practitioners belonged also to the ascetic mendicant orders" (1992: 163). However much the scientific revolution may have changed perceptions of the natural world, it only intensified patriarchal attitudes towards women (Merchant, 1980). Francis Bacon's slogan "knowledge is power" and his metaphors of the male scientist controlling female nature retained the main themes of ascetic patriarchy as the agenda for the new science (Keller, 1985), while Descartes built the dualisms of classical theology into the foundations of modern philosophy (Ruether, 1972). As science became institutionalized, those few "back doors" through which women could enter science—salons, guilds, and aristocratic courts—were closed in the name of "professionalization" (Schiebinger, 1989). Ruether summarizes the consequences:

> All the basic dualities—the alienation of the mind from the body; the alienation of the subjective self from the objective world; the subjective retreat of the individual, alienated from the social community; the domination or rejection of nature by spirit—these all have roots in the apocalyptic-Platonic religious heritage of classical Christianity. But the alienation of the masculine from the feminine is the primary sexual symbolism that sums up all these alienations. The psychic traits of intellectuality, transcendent spirit and autonomous will that were identified with the male left the woman with the contrary traits of bodiliness, sensuality and subjugation. Society, through the centuries, has in every way profoundly conditioned men and women to play out their lives and find their capacities within this basic antithesis. (1972: 115)

The values, language and modes of thinking of ascetic patriarchy have, through this inheritance, become the norms and discourse of science and technology. For those feminists who make this argument, though, the traditional spheres of men and women have been transvalued (Tong, 1989: 135). In opposition to patriarchy, they see what is taken to be "essentially feminine" as good.

Dichotomous value sets characterize men and women, these feminists argue (e.g., Greenbaum, 1987). Men are associated with words such as objective, reason, impersonal, rational, power and things. Words applied to women are just the opposite: subjective, feeling, personal, emotional, love, and people. This dichotomy, according to Greenbaum (1987: 3), parallels the dominant understanding of science and nature. The ideal of a rational, objective science and a disciplined, emotionally controlled masculinity have become intertwined, these theorists say, "science has been linked with male power and dominance, objectivity and neutrality, coldness of feelings, and destruction of nature and humanity" (Elkjær, 1986: 51). Furthermore, the links between science and patriarchy have become deeply held personal values for many scientists and engineers. Consequently, any challenge to the values of science threatens many men's identity. When their identity is threatened, they respond aggressively. According to Høyrup: "Traditionally, science and technology have been men's subjects, and men have much pride in excelling in these fields. Therefore, many men are not interested in getting women into these subjects, then they feel their manhood reduced, if women also excel in these 'rational' subjects" (1986: 92). Thus the resistance to women in precisely those disciplines which most pride themselves on their dedication to reason and truth (cf. Harding, 1986; Keller, 1985).

This argument differs from Romanticism (and some other forms of feminism) in that gender itself is understood as having been constructed, that is, the different value sets and cognitive styles characteristic of men and women are not innate or biologically determined but are the product of culture. Patriarchy is not inevitable, they argue, but it is powerful, shaping science and technology in the interests and image of men. So, they say, gender is ultimately a social construction, but once constructed gender identities are embodied and become normative. They determine the direction of science and technology. In effect, this variety of feminism has substituted a form of gender determinism for technological determinism.

Authoritative Knowledge

The strongest example of the essentialist argument is over what Suchman and Jordan call *authoritative knowledge*, which is defined as "that knowledge taken to be legitimate, consequential, official, worthy of discussion and useful for justifying actions by people engaged in accomplishing a given task" (1989: 153). It governs discourse in that it defines what are the appropriate ways to think and argue and act. A number of studies postulate a dichotomy between a masculine authoritative knowledge and the knowledge of women. This dichotomy is variously described as between rule-based or holistic learning (Brecher, 1989), field-dependent and field-independent cognitive styles (Fowler and Murray, 1987), the head and the heart (Greenbaum, 1987), and technically limited rationality or responsible rationality (Verne, 1986). The most widely used distinction is between what Sherry Turkle called "hard mastery" and "soft mastery" (1984a, 1984b). In 1990, Turkle revised her analysis (Turkle and Papert, 1990). It is this revision I will use as a case study.

In a study of how students actually go about programming computers, Turkle and Papert discovered two cognitive styles. The first is the formal approach which emphasizes structure, abstract manipulation of symbols, hierarchy, and planning. It dominates the computer culture and forms its authoritative knowledge. It is also a form of knowledge strongly associated with men. In opposition to this is a style they call *bricolage.* The bricoleur is more concrete and tactile than abstract, more intimately involved than distant and objective. They describe the differences:

> The bricoleur resembles the painter who stands back between brushstrokes, looks at the canvas, and only after this contemplation, decides what to do next. For planners, mistakes are missteps; for bricoleurs they are the essence of a navigation by mid-course corrections. For planners, a program is an instrument for premeditated control; bricoleurs have goals, but set out to realize them in the spirit of a collaborative venture with the machine. For planners, getting a program to work is like "saying one's piece"; for bricoleurs it is more like a conversation than a monologue. In cooking, this would be the style of those who do not follow recipes and instead make a series of decisions according to taste. While hierarchy and abstraction are valued by the structured programmers' planner's aesthetic, bricoleur programmers prefer negotiation and rearrangement of their materials. (1990: 136)

Most (but by no means all) bricoleurs are female. These two cognitive styles are thus *gender-linked* but not *gender-determined.* Some men are alienated from the authoritative style of the computer culture while many women work well within it (1990: 150).

As different as these two cognitive styles may be, Turkle and Papert emphasize that *both* produce excellent results. Computers, with their capabilities in music, text, graphics, and sound can accommodate a variety of cognitive styles. Icons and graphical user interfaces are particularly amenable to bricoleurs.

The problem is that "although the computer as an expressive medium supports epistemological pluralism, the computer culture often does not" (1990: 132). The problem does not lie in the technical aspects of technology practice (the computers themselves) but in the cultural. Because the formal approach, as the authoritative knowledge, claims to be the only right way to do computing, bricoleurs are often caught in a conflict. This has two dimensions. First is the immediate conflict between the bricoleur's personal style of knowing and acting and what is expected of them. Most deal with this conflict by withdrawing, forcing themselves to conform, or "faking it." Complicating this is a second dimension, its overdetermination by gender.

Our society describes the authoritative knowledge of computing as male because it is associated with (indeed, it is derived from) "a construction of science that stresses aggression, domination, and competition" (1990: 150). It is no accident that the dominant values of science and of most men are the same. Turkle and Papert claim that from its beginning, science and male knowledge have been linked:

> From its very foundations, science has defined its way of knowing in a gender-based language. Francis Bacon's image of the (male) scientist putting the (female) nature "on the rack," underscores the way objectivity has been constructed not only in terms of the distance of the knower from nature but also in terms of an aggressive relationship toward it (or rather toward her). From its very foundations, objectivity in science has been engaged with the language of power, not only over nature but over people and organizations as well. Such associations have spread beyond professional scientific communities; aggression has become part of a widespread cultural understanding of what it means to behave in a scientific way. Its methods are expected to involve "demolishing" an argument and "knocking it down" to size. Here the object of the blows is not a female nature but a male scientific opponent. Science is first a rape, then a duel. The traditional discourse of computation has not been exempt from these connotations. (1990: 150-51)

Women in computing are thus caught in a conflict between the social definition of their work (embodied in the authoritative knowledge) and their identity as women. So on top of whatever conflict a bricoleur may face with peers, co-workers, teachers, and supervisors, "the computer culture alienates by putting one in conflict with oneself" (1990: 151). Confronted with such conflicts, it is hardly surprising many women turn away from computing. Turkle and Papert call for epistemological pluralism to counteract the intolerance of the authoritative knowledge, recognizing that the computer culture itself will have to be profoundly transformed.

The debate over authoritative knowledge is perhaps the strongest form of the essentialist argument. It sees the origin of differences in men's and women's participation in computing not in who they *are* but in how they *know*. Computers are seen as essentially masculine because male values, discourse, and forms of knowing have formed the practice of computing. This is a more sophisticated argument than the Romanticism of some varieties of essentialism, which frequently spoke of an "eternal feminine" in opposition to an abstract but threatening technology, but it still substitutes gender determinism for technological determinism.

Turkle and Papert, however, make a crucial distinction (which many others in the debate do not) when they imply that their categories are gender-linked but not gender-determined. To say something is gender-linked means that gender is associated with, but not inherent in, the phenomenon. It avoids rigid dichotomies while still emphasizing the central role of gender in the computer culture. This distinction saves their theory from simple gender determinism.

This insight has two implications which expose contradictions in the essentialist argument. First, if many men are bricoleurs, while some women thrive in a "masculine" computer culture, a simple gender dichotomy loses its explanatory power. More is required by way of explanation than an account of one or the other sets of values or cognitive styles. In other words, value sets and cognitive styles may be necessary to understand the gender-typing of technology but it is not sufficient for an explanation. Research would have to find what part computers actually play in the construction of gender identities.[4]

Second, if computers are a masculine machine because the values of the computer culture are those of men, how did men impose their values on computing, or, to be more precise, in what way do gender identities construct the way computers are perceived at the levels of language and discourse? If, as some argue, that the values of science and technology (in general) are those of men, then all technologies should be masculine. But this is not the case—typewriters, sewing machines, stoves, and washing machines (to name just a few) are all identified with women. What essentialist arguments cannot answer is how technology practices become appropriated by one gender or the other. They have an inadequate account of the application of power because they pay insufficient attention to the organizational aspect of technology practice.

The Process of Appropriation

The second argument over why artifacts are gendered believes the problem is that computers are not so much essentially masculine as that they have been appropriated by men. The emphasis in these arguments is on the power relationships of the organizational aspect of technology practice.[5]

Any technology is practiced by diverse groups of people, whom Wiebe Bijker and Trevor Pinch dubbed *relevant social groups* (1987), who may be the inventors, developers, sponsors, manufacturers, or users of that technology. Every technical innovation is surrounded by a web of interests—the developers and those who pay them, their rivals, who may be either trying to develop another version of the innovation or sponsoring an alternative technology, businesses which stand to profit from, or are threatened by, the innovation, policy makers who may wish to regulate or tax it, and many groups of potential users who may constitute its market.

---

4 Which Turkle has pursued in her most recent work (1995).

5 The account I am giving here is a variety of constructivism called the Social Construction of Technology (SCOT). There are other constructivist accounts. See, for example, the essays in: Bijker and Law, 1992; Bijker, Hughes and Pinch, 1987; Kramarae, 1988; and Mackenzie and Wajcman, 1985; and books by Collins and Pinch, 1993; and Latour, 1987, 1996.

The meaning of any technology is created by a relevant social group, which will, as we saw in the second chapter, construct a technological frame based upon their own interests and how they perceive the technology as solving (or adding to) their problems. As long as the debate over a technology's meaning remains open, all the relevant social groups contend to impose their definition and understanding. Bijker (1993, 1995) sees three possible configurations of frames in such a situation. When no single group is dominant there is likely to be a multiplicity of frames, when there are two or three entrenched groups, each will pursue its own frame, and when only one group is dominant there will be a single hegemonic frame. Each group in this conflict will follow one or more strategies, such as trying to enrol additional groups into their frame or redefining the problem addressed by their frame in order to create a broader constituency (Bijker, 1995). Should one of these groups be successful, we can say they have *appropriated* (Elkjær, 1989) that technology, that is, their definitions, theories, metaphors and so on which make up the technological frame have prevailed. When this happens, it is quite likely that *stabilization* and *closure* (Bijker, 1993, 1995) will occur, which is to say, the victorious social group will impose its definition as the "universal" meaning of the technology and define its practices as "normal" and a "consensus."

In the first configuration, there is likely to be a low degree of stabilization and closure of debate is less likely or, if achieved, likely to be short-lived and precarious as innovation continues. The third configuration, on the other hand, is marked by a high degree of stabilization and firmly entrenched closure. Further innovation in the affected technology is likely to be slower and rather narrowly focused. It is quite possible for a single technology to pass through all three configurations over time. What was obviously a social construction while the debate was still open appears to be natural, inevitable, and the only way things could be as the debate is forgotten (Marvin, 1986). For example, in a famous study done by Pinch and Bijker (1987; Bijker, 1995), the early high-wheeled bicycle was constructed by "young men of means and nerve" (1987: 34) as the "macho machine" while to older men it was the "unsafe machine" and to women it was both unsafe and immodest. Each constructed a technological frame which gave meaning to the technology and through which they attempted to appropriate it. In the case of the bicycle, women and older men were successful, and bicycles were redefined and redesigned into the low-wheeled "safety" bicycle. A century later, the safety bicycle is universal and taken-for-granted, it is "the way bicycles are supposed to be," while the high-wheeled machine is remembered only as a curious antique.

Once a social group has successfully appropriated a technology, its technological frame becomes the "objective" and "universal" definition of that technology's practice. Other groups are free to participate but they do so *on the terms* of the dominant group. In Bijker's terms, a group may have a high or low degree of *inclusion* into the technological frame. For those with a high degree of inclusion the "artifact is *unambiguous* and *constrains* action, but it is also highly *differentiated* and enables them to do many things" (1995: 283). Conversely, "for actors with low inclusion, the artifact presents a 'take it or leave it' decision" (1995: 284). For example, men decide what the practice of computing involves, women may then participate in *that practice* (but not in another). Who wins the struggle to appropriate a technology is not predetermined, but not all groups are equal in the fight. New technology is not dropped onto a *tabula rasa*—there is never a moment in the life cycle of a technology when it is not always already immersed in politics and values. It becomes part of the ongoing power relations of society. Some groups have more resources and allies available to them by virtue of their position in society, and while it is not inevitable that their frames will prevail, the times when they do not are the exception rather than the rule.

This is because, in most cases, the construction and appropriation of a technological frame is part of what Dorothy Smith (1987, 1990) calls the *relations of ruling*. She defines the relations of ruling as:

> An ideologically structured mode of action—images, vocabularies, concepts, abstract terms of knowledge [which] are integral to the practice of power, to getting things done. Further, the ways in which we think about ourselves and one another and about our society—our images of how we should look, our homes, our lives, even our inner worlds—are given shape and distributed by the specialized work of people in universities and schools, in television, radio and newspapers, in advertising agencies, in book publishing and other organizations forming the "ideological apparatuses" of society. (1987: 17)

As one aspect of the relations of ruling, technological frames are a particular way of constructing reality which orders experience. They abstract from lived experience to create a textually mediated presentation of "reality" which, once established, becomes the filter through which experience, and even identity, are defined and made meaningful.[6] Furthermore, they appear to be "objective" and "universal," that is, simply the "way things are."

The point Dorothy Smith makes is that the relations of ruling abstracts the experience of one social group—men—and translates their particular experience into a universal. The lives of *men* becomes the universal *man*. Women disappear into this "universal," their own lived experience not only discounted but filtered through a normative standard

---

[6] See chapter five for a further discussion of how this occurs.

which is alien and at times hostile. The world we perceive is thus structured and ordered, the categories we use to understand reality given, the language of our discourse selected.

Thus the development of any technology occurs in the context of a struggle over that technology's essential definition and its appropriation. When a group is successful in appropriating a technology, stabilization and closure occur, that is, active debate ceases. The winners of the struggle define the technical culture, and become the experts upon which everyone else is dependent (Doniol-Shaw, 1989). The technological frame of the victorious group now appears to be the universal meaning of the technology—it becomes "reality," a black box that can only be reopened with difficulty.

Making Computers Masculine

There are many strategies followed in the struggle to appropriate a technology. Some of the most effective include various socialization techniques and gatekeeping devices. Socialization techniques include all the various ways people have of learning, from formal education to personal interactions with family and peers. Gatekeeping devices are socially constucted barriers and obstacles to entrance or participation in a group. There are a wide variety of these devices. Some are informal procedures such as jargon unintelligible to the outsider or initiation rites (some engineering schools' wet T-shirt contests and sexist school newspapers, for instance). These create images and expectations which discourage all but the most determined from choosing to participate. At the extreme end of the continuum are formal exclusionary measures such as quotas or blacklists. A case study will show how these techniques are used to help white men appropriate computing.

One of the socialization and gatekeeping techniques which has not been adequately studied is computer advertising. There is a substantial body of literature on the potent role advertising plays in gender socialization (Belknap and Leonard, 1991; Bretl and Cantor, 1988; Downs and Harrison, 1985; Gould, 1987; Sullivan and O'Connor, 1988), but only a few studies have looked at computer advertising. The first of these was a content analysis of computer advertising by Demetrulias and Rosenthal (1985), which discovered the prevalence of cultural stereotypes. The most common type of ad portrayed a single white male. Males were commonly portrayed in athletic settings or—ubiquitously dressed in business suits—engaged in serious work. Females, when shown at all, were usually surrounded by "feminine" images or situations (e.g., in the kitchen, as a secretary) and almost always in a subordinate or peripheral position.

People of colour were even less visible and portrayed just as poorly. They concluded:

> Cultural stereotypes and other societal forces perpetuate beliefs and attitudes that discourage women and racial minorities from fully participating in the world of computers. Microcomputers advertisements are one source, perhaps a major source, of sex-role socialization, and this study found that they reinforce traditional gender and racial stereotypes. (1985: 95)

Since 1985 there have been considerable changes in the computer industry, in computer magazines, and in how computers are marketed. Unfortunately, these changes do not extend to how women and people of colour are portrayed.

Today the computer market is fragmented. Very few of the general interest computer magazines surveyed by Demetrulias and Rosenthal exist today.[7] They have been replaced by more specialized magazines targeting more clearly defined market niches. Fragmented markets mean fragmented images. It is risky, therefore, to generalize about "computer advertising" apart from the images portrayed to each niche. As we will see, while women and people of colour are underrepresented in every niche, the quality of their portrayal varies considerably. Differing images between niches may give us some clues how the dominant group creates and maintains its position in the computer culture.

In this case study I examined the December 1992 or January 1993 issues of twenty computer magazines and analyzed the computer advertisements. The magazines chosen were those readily available on the newsstands or at computer stores, to encompass the images presented to the public. Academic and technical journals were excluded. The aim was to capture a "snapshot" of current images, rather than to systematically cover every possible niche or follow trends over time.

An advertisement was considered a "computer ad" if it was for any item of hardware, software, peripheral device, or other computer-related product, publication, or service. Non-computer ads[8] and classified ads were excluded. Since the aim of this case study is to find out how each magazine portrays people, rather than how advertising in general does, no attempt was made to exclude ads found in more than one magazine[9] (there are an appreciable number of ads duplicated within a niche but very few

---

7 Only two, *Byte* and *Compute,* were part of both studies.

8 There were very few of these ads. No magazine had more than seven and most had none at all.

9 Here my approach differs markedly from that of Demetrulias and Rosenthal. Additionally, their aim was to survey images presented to educators, not the general public. Because of these differences, this work cannot be considered a replication of the 1985 study, and comparisons should only be made with caution.

are duplicated between niches). In the case of multipage spreads for mail-order firms offering a variety of products from many companies, each product was counted as a separate ad if it stood out and was clearly identified in its own right. Ads that simply listed multiple products were counted as one ad for the mail-order firm. Multiple page spreads for one company or product were counted as one ad.

I proceeded through three iterations. First, I identified those ads which portrayed people and did a frequency count of the numbers of men and women and of people of colour presented. Serious overrepresentation by white males would show computing as a "normal" activity for them but not for others. Second, I analyzed the quality of the images of women and people of colour. Racist and sexist ads both reinforce stereotypes held by those within the computer culture and produce an image of that culture to those outside it. Third, I looked at themes the ads aim at men. One theme stands out—computers are a source of power. Taken together, these three iterations should give a fairly clear indication of how advertising helps to shape and reproduce the racial and gender types of the technology, discouraging women and people of colour from choosing to participate.

First Iteration: Frequency

The fragmentation of the computer market has led to the rise of specialist magazines, each appealing to a particular market niche. Each of the magazines studied here was identified by niche (see Table 3.1). *Compute* is one of the few general interest computer magazines left. Business-oriented magazines account for the largest bloc, but they are in turn specialized by machine—PC, Windows, or Macintosh. These tend to be thick journals consisting almost entirely of advertisements and product reviews. One magazine catering to small business was found. The programmers niche incorporates those magazines aimed at computer professionals. The home hobbyist niche is a collection of relatively small magazines directed to the home computer market, each specializing in a different machine. Game magazines are divided into two subniches, computer games, those which are played on a microcomputer, or video games, such as Nintendo or arcade games (a number of games are found in both formats). *Interaction* is the house organ of Sierra On-Line and its format differs somewhat from the other two, which are little more than compendiums of game reviews and advertisements.

Most ads (other than those for games) are fairly straightforward. They show a picture of the product and describe its price and features. A significant percentage, however, portray people (see Table 3.1), ranging from 58.48 percent of total computer ads in game magazines, to 22.73 percent for the business niche, to 18.39 percent in the programmers niche, to a low of 13.56 percent of home hobbyist ads (although this last group varies considerably from one magazine to another). With the exception of two of the game magazines, the person most frequently portrayed by these ads is a single male. For many magazines single men are found in more than half of all the ads portraying people. The next largest category (with the exception of *Byte* and two game magazines) is ads showing both men and women, followed by an all male group.[10] *Computer Gaming World* and *Electronic Gaming Monthly*, whose audiences are primarily adolescents and young adults, have the all male group as their largest classification. There are fewer ads showing a single woman than those showing single men or mixed groups in every magazine and five magazines, all from the home hobbyist and games niches, show no single women at all. Even more rare is an ad featuring a group of women—only nine ads scattered over six magazines.

Table 3.1

Frequency Distribution of Computer Ads Portraying People

Magazine	Computer Ads			Male Group		Female Group		Mixed Group		Single Man		Single Woman	
	Total	People	% People	N	%	N	%	N	%	N	%	N	%
GENERAL													
Compute	152	59	38.82	8	13.56	1	1.69	17	28.81	22	37.29	11	18.64
BUSINESS - PC													
PC Magazine	298	47	15.70	7	14.90	0		10	21.30	21	44.70	9	19.10
PC Sources	227	39	17.20	6	15.40	0		12	30.77	15	38.46	6	15.38
PC Computing	252	69	27.38	7	10.14	0		16	23.19	36	52.17	10	14.49
PC World	285	73	25.61	8	10.96	0		10	13.70	49	67.12	6	8.22
BUSINESS - WINDOWS													
Windows User	97	25	25.80	1	4.00	0		7	28.00	13	52.00	4	16.00
BUSINESS - MACINTOSH													
MacWorld	205	46	22.44	3	6.52	1	2.17	16	34.78	19	41.30	7	15.22
MacUser	290	77	26.55	5	6.49	3	3.90	16	20.78	39	50.65	14	18.18
SMALL BUSINESS													
Home Office Computing	67	24	35.80	1	4.20	0		7	29.20	12	50.00	4	16.70
PROGRAMMERS													
Byte	216	44	20.40	7	16.00	1	2.30	4	9.00	31	70.50	1	2.30
Unix World	148	33	22.30	3	9.00	0		4	12.10	24	72.70	2	6.00
Computer Language	120	12	10.00	2	16.67	0		4	33.33	5	41.67	1	8.33
HOME HOBBYIST													
PC Novice	15	9	60.00	1	11.11	0		5	55.56	3	33.33	0	
Mac Home Journal	51	7	33.30	0		0		3	17.67	10	58.80	3	17.60
Run (Commodore)	29	4	13.80	0		0		1	25.00	3	75.00	0	
InCider/ A+ (Apple)	51	6	11.76	1	16.67	0		2	33.33	3	50.00	0	
Amiga World	149	14	9.40	0		2	14.29	3	21.43	6	42.86	3	21.43
COMPUTER GAMES													
Computer Gaming World	126	63	50.00	31	49.20	0		13	20.63	18	28.57	1	1.58
Interaction	31	14	45.16	2	14.00	0		5	36.00	7	50.00	0	
VIDEO GAMES													
Electronic Gaming Monthly	120	85	70.83	49	57.65	1	1.18	13	15.29	22	25.88	0	

---

[10] Note that a considerable number of the ads classified as showing a "group" might better be termed "multiple individuals."

Table 3.2

Gender Division of Computer Ads Portraying People

Magazine	Men Only		Women Only		Mixed Group		Ratio	Overall Ratio
	N	%	N	%	N	%	M:F	M:F
GENERAL								
Compute	30	50.85	12	20.34	17	28.81	1.79:1	3.04:1
BUSINESS - PC								
PC Magazine	28	59.50	9	19.15	10	21.30	2.08:1	3.48:1
PC Sources	21	53.80	6	15.38	12	30.77	2.0:1	2.76:1
PC Computing	43	62.32	10	14.49	16	23.19	3.0:1	3.71:1
PC World	57	78.08	6	8.22	10	13.70	.92:1*	2.75:1
BUSINESS - WINDOWS								
Windows User	14	56.00	4	16.00	7	28.00	1.63:1	3.25:1
BUSINESS - MACINTOSH								
MacWorld	22	47.83	8	17.39	16	34.78	2.0:1	2.47:1
MacUser	44	57.14	17	22.08	16	20.78	1.93:1	2.13:1
SMALL BUSINESS								
Home Office Computing	13	54.20	4	16.67	7	29.20	.9:1	1.64:1
PROGRAMMERS								
Byte	38	86.36	2	4.55	4	9.09	3.33:1	6.82:1
Unix World	27	81.80	2	6.06	4	12.10	2.0:1	7.5:1
Computer Language	7	58.33	1	8.33	4	33.33	6.57:1	7.5:1
HOME HOBBYIST								
PC Novice	4	44.44	0		5	55.56	1.0:1	2.4:1
Mac Home Journal	10	58.82	3	17.65	3	17.67	1.75:1	4.25:1
Run (Commodore)	3	75.00	0		1	25.00	.5:1	2:01
InCider/ A+ (Apple)	4	66.67	0		2	33.33	2.0:1	3.29:1
Amiga World	6	42.86	5	35.71	3	21.43	1.33:1	0.83:1
COMPUTER GAMES								
Computer Gaming World	49	77.78	1	1.59	13	20.63	2.05:1	10.07:1
Interaction	9	64.29	0	- -	5	35.71	1.85:1	3.83:1
VIDEO GAMES								
Electronic Gaming Monthly	71	83.53	1	1.18	13	15.29	4.83:1	21.42:1

* 1.87:1

Table 3.2 further clarifies the gender division of these ads. For all but four magazines, *Compute, MacWorld, PC Novice,* and *Amiga World,* ads depicting only men are the overwhelming majority. Only in *Compute, MacUser*, and *Amiga World* does more than one ad in five display women only. Four magazines have no ads depicting only women, and two others have only one each. The remaining ads show a mixed group of men and women, but their handling is far from equal. When the numbers of individuals in these ads are counted and compared, the ratio of men to women is 2:1 or higher in half the magazines. In *Computer Language* the ratio is 6.57:1, more than offsetting its relatively lower percentage (for its niche) of men only ads. In only four magazines was the ratio of men to women 1:1 or less, and two of these are anomalies. *Run* has only one ad portraying women, depicting two women and one man. *PC World*'s low ratio is caused by one game ad that drew a large number of very small faces, mostly women. If that one ad is excluded, the ratio for remaining ads jumps to 1.87:1.

The final column on Table 3.2 compares the total number of men and women (not just in mixed group ads). In only two magazines did the ratio of men to women fall below 2:1. *Amiga World* had a ratio of 0.83:1 (its ads showed ten men and twelve women), and *Home Office Computing* depicted 1.64 men for every woman. By far the worst was *Electronic Gaming Monthly* at 21.42:1.

So, based simply upon the frequency with which men and women are portrayed in computer ads, the message is sent that computing is a "man's world." Figure 3.1 graphically demonstrates the effect. While women are underrepresented in every niche, their exclusion from the programmers and games niches is particularly pronounced. Since magazines in the programmers niche cater to computer professionals—the "insiders" of the field—and games are usually the first contact many adolescents have with computing, women's underrepresentation here is likely to have a particularly strong effect on the computer culture.

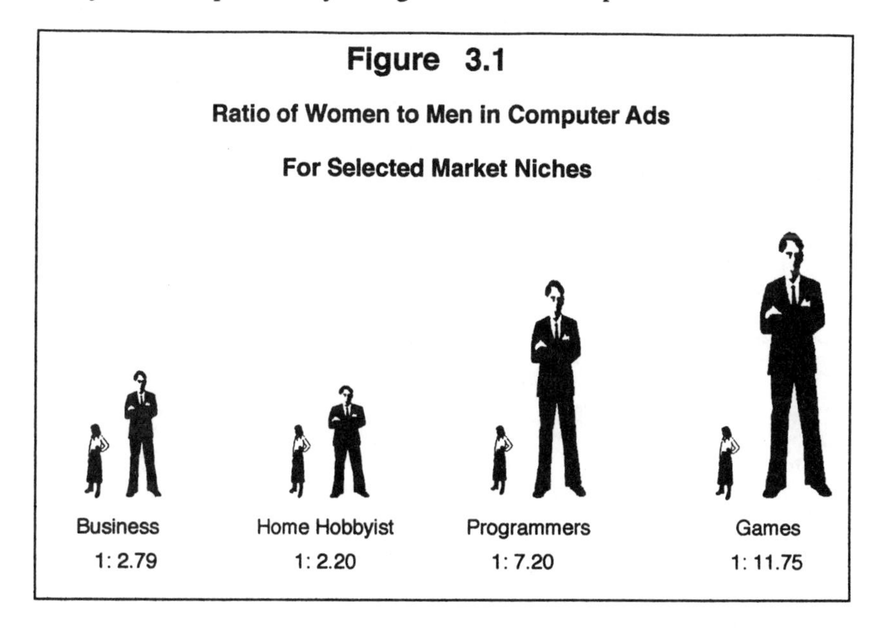

Figure 3.1

Ratio of Women to Men in Computer Ads

For Selected Market Niches

If women are underrepresented in computer ads, much the same can be said for people of colour. As Table 3.3 delineates, in only four magazines do ads depicting people of colour exceed 10 percent of those ads portraying people. In three of them, this percentage is obtained with an N of only one or two ads. A more accurate measure of representation is what percentages of all individuals in computer ads are people of colour. In only three magazines are they more than 10 percent of all people portrayed. In five magazines *all* the people pictured are white. The unusually large number of men of colour in *Computer Gaming World* and *Electronic Gaming Monthly* is produced, in large part, by ads showing sports teams or groups of soldiers. Women of colour are virtually absent from computer advertising. Although in *Interaction* and a few business and home hobbyist magazines they achieve parity, the numbers involved are extremely small. By any measure, then, people of colour are seriously underrepresented in every niche. Computer advertising thus depicts computing as an activity primarily engaged in by white men.

Table 3.3

People of Colour in Computer Ads Portraying People

Magazine	Number of Ads		% of all	Gender Division			
	N	%	people	N Male	% Male	N Female	% Female
GENERAL							
Compute	1	1.69	1.05	2	100	0	
BUSINESS - PC							
PC Magazine	3	6.30	4.00	3	75	1	25
PC Sources	1	2.56	4.00	1	33.33	2	66.67
PC Computing	6	8.70	5.00	7	87.5	1	12.5
PC World	4	5.48	2.42	2	50	2	50
BUSINESS - WINDOWS							
Windows User	2	8.00	3.92	2	100	0	
BUSINESS - MACINTOSH							
MacWorld	4	8.70	2.45	2	50	2	50
MacUser	3	3.90	1.84	3	50	3	50
SMALL BUSINESS							
Home Office Computing	0		0	0		0	
PROGRAMMERS							
Byte	2	4.50	2.33	2	100	0	
Unix World	0		0	0		0	
Computer Language	2	16.67	11.76	8	100	0	
HOME HOBBYIST							
PC Novice	0		0	0		0	
Mac Home Journal	1	5.90	9.52	1	50	1	50
Run (Commodore)	0		0	0		0	
InCider/ A+ (Apple)	1	16.67	3.33	0		1	100
Amiga World	0		0	0		0	
COMPUTER GAMES							
Computer Gaming World	4	6.35	11.29	54	98	1	2
Interaction	2	14.29	9.89	5	55.56	4	44.44
VIDEO GAMES							
Electronic Gaming Monthly	10	11.76	10.29	60	100	0	

Second Iteration: Quality

Presentation frequencies alone are not an adequate measure, however. Unless the quality of the portrayal is considered, frequency may prove deceptive. *PC World*, for example, has relatively fewer ads showing single women than other journals in its niche, but this is in part the result of *not* carrying computer pornography ads. To measure the quality of women's representation, I rated each ad in which a woman appears on the following scale:

> 1= overtly sexist (women ridiculed or portrayed as sex objects)
> 2= negative image (women in traditional roles or subordinate to men)
> 3= neutral
> 4= positive image (women in non-traditional roles)
> 5= very positive (women in positions of power and authority)

Thus an ad showing a woman as a secretary or a mother, or taking direction from a man, would be rated two, as a computer programmer would be rated four, as an entrepreneur or corporate executive would be rated five. Ads showing a woman's face or a woman at a computer, if no other cues were given as to her role, were rated three. Ads depicting women who were scantily clad, who were portrayed as giving a sexual "come on," who were made to look ridiculous, or whose features were distorted (common in ads for morphing software) were rated one. The resulting index for each magazine is displayed in Figure 3.2. A rating for a magazine lower than three should be considered a negative portrayal of women, lower than two as extremely negative.

As Figure 3.2 exhibits, only one magazine, *Home Office Computing,* can be considered to present a positive image of women overall, largely on the strength of a few ads featuring individual women entrepreneurs. *PC World* had a positive rating for single women (the absence of pornography ads raised it above others in its niche), while *Run, InCider/A+* and *Unix World* (for single women) presented neutral images on the basis of only one or two ads each. All other magazines were negative. While a number of ads pictured women engineers or executives, they were considerably outnumbered by images of secretaries and sex objects. In nine magazines the score for ads showing a single woman was lower than the rating for all ads featuring women. In other words, single women were portayed more negatively than women together with men in nearly half the magazines. *Amiga World*, which was the most equal in the numbers of women presented, scored low on the quality of their portrayal, largely because of graphics software ads depicting women as sex objects or with their features distorted (men were never deformed in this way).

The games magazines were extremely negative. *Electronic Gaming Monthly*, with a rating of 1.14, was by far the worst. Nearly all

Figure 3.2

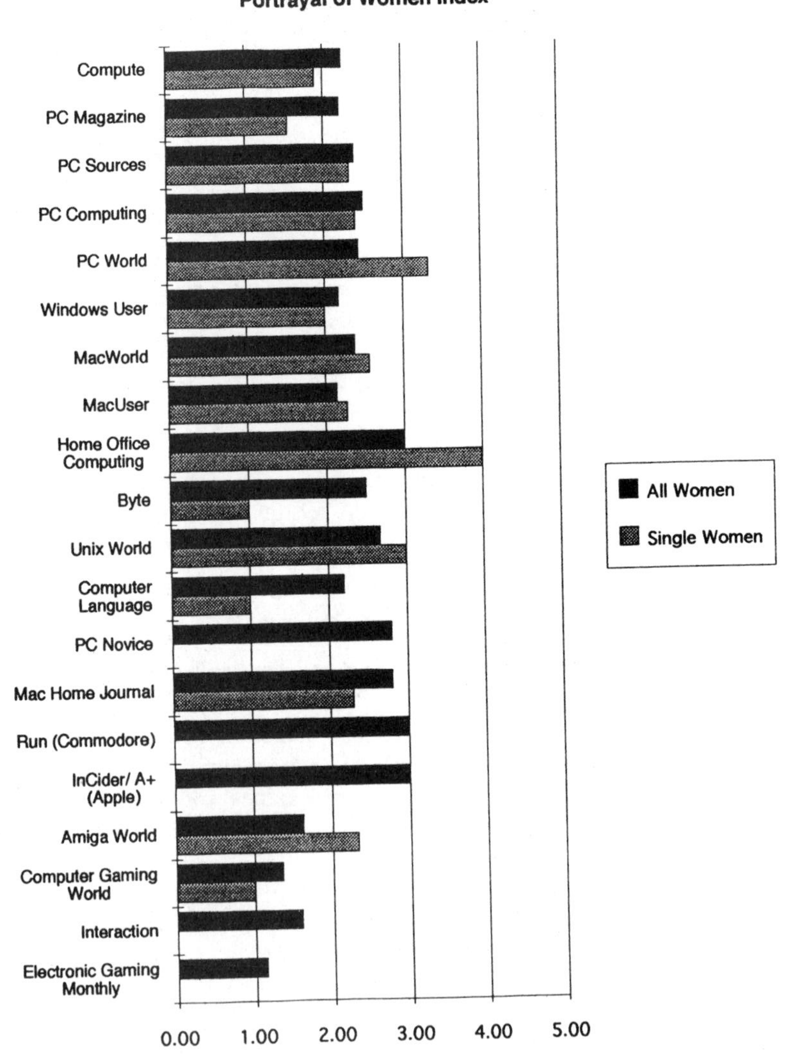

ads depicting women in this niche (and many computer game ads in the others) can only be described as misogynist. Women are almost invariably large-breasted and revealingly clothed. Many are victims. Most of these games are violent, and for some the violence is explicitly directed towards women.[11] Again, the particular significance of this niche lies in their youthful clientele and the prominent place games have in socialization into computing.

The quality of the portrayal of people of colour in these ads is decidedly mixed. Outside the games niches, virtually all people of colour are shown as part of a mixed group with whites, usually also a mixed group of men and women. The most common images in these niches are either children or business people/computer professionals. On the whole these images are neutral or favourable, although one ad run in several business magazines depicts a single black male in a stereotypically "tribal" costume. In the games magazines, people of colour tend to be shown as either athletes or soldiers. While many of these ads feature sports heroes, some are openly racist. For example, one proclaims; "DON'T GET MAD...GET EVEN! Your chance to get your own back!" for a game entitled Conquest of Japan[12] (*Computer Gaming World*, Dec. 1992: 51). Another draws a male Native American as a crudely stereotypical "cigar store Indian." Yet another mocks surrendering Iraqi soldiers. More serious, in terms of quality, is the extremely narrow range of roles allotted to people of colour in these ads, particularly the stereotype of the Afro-American athlete. And while these ads may picture people of colour as characters *in* a game, none show a person of colour actually *playing* a computer game.

Third Iteration: Power

Our final concern is the themes which computer ads aim at men. Running through all these ads is a persistent theme—computers are a source of power. "Power" is one of the most frequently used words in discourse about computers and, indeed, many products incorporate the term into their names (e.g., Apple's Powerbook computer). Many ads give the message, directly or subtlety, that their product is a means to potency. Figure 3.3 displays the frequency of the power theme. The percentages are of all computer ads in a magazine (many ads convey the power theme without picturing people). This power theme is found least frequently in home

---

[11] The worst ad in this regard is one for The Combatribes by Nintendo (*Electronic Gaming Monthly*, Jan. 1993: 53). In the form of a full-page comic strip, the ad depicts a beating administered to a well-endowed blonde in a skimpy dress, while the text describes the object of the game as: "Your Mission: destroy Martha Splatterhead—former Combatribe-babe turned Gotham gang-queen."

[12] Ironically, the game itself appears to be a simulation of the civil wars of medieval Japan. The sales pitch seems to be at variance with the content of the product.

hobbyist magazines, makes up a significant percentage of the ads in the business niche, and dominates games advertising.[13]

The message directed at adolescents is neither subtle nor ambiguous. The image of masculinity presented by many game ads is

Figure 3.3

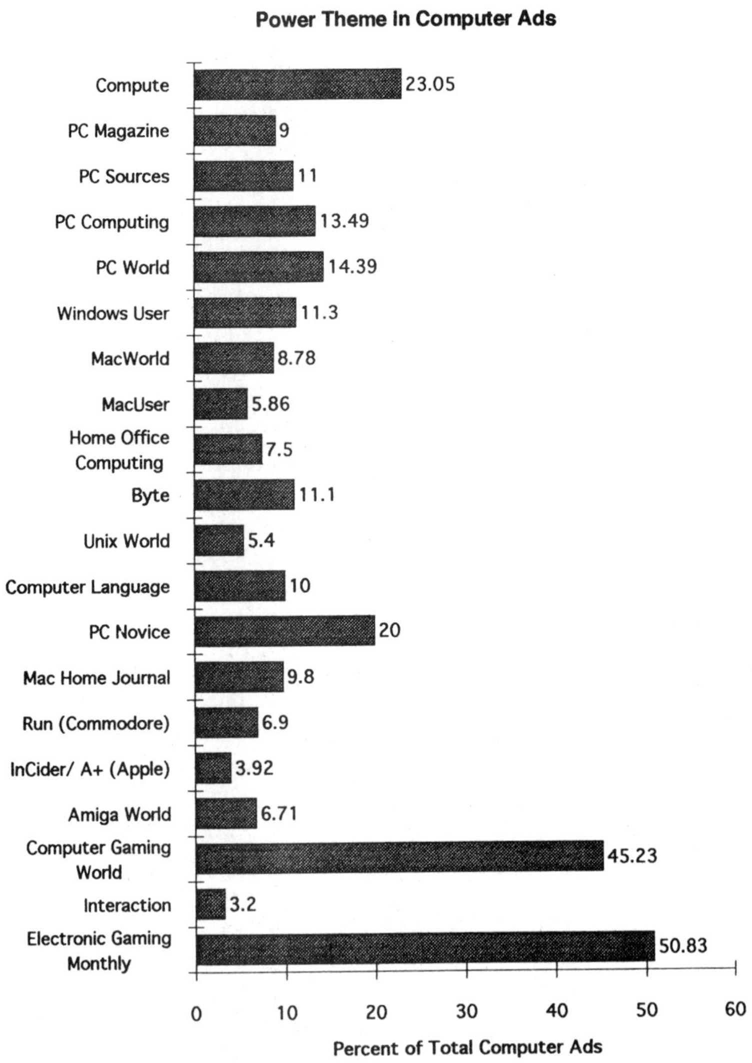

Power Theme in Computer Ads

Magazine	Percent of Total Computer Ads
Compute	23.05
PC Magazine	9
PC Sources	11
PC Computing	13.49
PC World	14.39
Windows User	11.3
MacWorld	8.78
MacUser	5.86
Home Office Computing	7.5
Byte	11.1
Unix World	5.4
Computer Language	10
PC Novice	20
Mac Home Journal	9.8
Run (Commodore)	6.9
InCider/ A+ (Apple)	3.92
Amiga World	6.71
Computer Gaming World	45.23
Interaction	3.2
Electronic Gaming Monthly	50.83

0 10 20 30 40 50 60

Percent of Total Computer Ads

---

13 With the notable exception of ads by Sierra, as is reflected in the figures for *Interaction*.

particularly crude. In the majority of game ads the options for men are athletics or war. A typical picture is a well-muscled, bare-chested "he-man," teeth clenched in a scowl, engaged in an act of violence. The power theme links this icon of masculinity with purchase of the product. For example, one ad in *Electronic Gaming Monthly* reads: "If only I could skate, slam, rage, jam, thrash, scam and score like I do with the asciiPad SG, that'd be killer!" (Jan. 1993: 87). It concludes: "With this kind of control, you'll always land on your feet. Which is more than we can say about life. The asciiPad SG. There's nothing fair about it...It's how to win." This ad recites all the motifs which patriarchy directs to males—be competent, be in control, winning is all that counts—and offers a piece of technology as the means to power. And the message is repeated in ad after ad aimed at young white men.

As the target audience of computer ads grows older, the power theme becomes somewhat less obvious and is repeated less frequently, but it appears often enough to keep the message clear. The majority of computer users are pictured as men either doing productive work or engaged in competitive sports. Images of masculinity and computers are repeatedly linked in, for instance, ads showing pictures of weight lifters with texts such as "Attention Power Users!" or "Pump Up Your Productivity." Themes of competency and control are still emphasized, but now more often as something to be realized and maintained rather than just fantasized about, as in an ad that shows a dapper professional man with the text: "I couldn't find a powerful CASE tool that was affordable. So I designed one" (*Computer Language*, Jan. 1993: 32-33). Thus for both younger and mature males, computers are depicted as a means of buying potency.

Advertising as Appropriation

These three iterations all point to a pattern of appropriation of computing by white men. As one aspect of computers' technological frame, the advertisements we have been examining play an important role in reflexively creating and reproducing the computer culture, and with that, its racial- and gender-types. They do this in three ways.

First, computer advertising proclaims that the norms and values of white men and of the computer culture are the same. Frequency of appearance establishes who "should" be doing computing. By far the most common depiction of a person in these ads is a single white man, and men outnumber women in every niche. Patriarchal values are consistently expressed. The message in many ads is that computers are the way to be competent, competitive, and in control. They are a source of power. The ads communicate that it is "normal" and acceptable for white men to do computing.

Second, these ads convey the message that if white men are the norm, women and people of colour are the exception. Women and people of colour are seriously underrepresented, their appearance in computer ads falling well below their proportion of both the population and the labour force. When they *are* portrayed it is most often as part of a mixed group in which white males are the majority. Women are infrequently shown alone and almost never in a group with other women. Furthermore, the content of these ads is frequently sexist and occasionally racist. Only one magazine of the twenty surveyed produced a positive image of women overall.[14] By tacitly saying to women and people of colour "you are an exception, you don't belong in the computer culture," these ads discourage them from wanting to participate. By tacitly saying to white men "your attitudes and understandings are universal and natural," these ads reinforce and legitimate behaviours and practices which tend to exclude women and people of colour.

Third, having played an important role in creating computing's racial- and gender-types, these ads aid in their reproduction. The different patterns between niches suggests the role these ads play in socialization into the computer culture. Magazines in the games niches are aimed at adolescents and young adults. These magazines have the highest percentages of ads portraying people, the largest number of ads showing men in groups, the highest ratio of men to women in the ads, the most negative portrayal of women, and the greatest incidence of racial stereotypes. They also have the strongest power theme. Is it coincidence that all this occurs in magazines aimed at people just entering the computer culture? In the other niches the pattern is reproduced, albeit more subtlety and with variations depending upon the target audience, as a reminder of the computer culture's norms and values.

In 1985 Demetrulias and Rosenthal concluded their study by saying of the bias against women and people of colour which they found: "While such stereotypes may not be intentional on the part of advertisers, the result of these images may be to cancel the extensive efforts of educators and others to create an environment of computer equity" (1985: 95). Since then, computer advertising has not changed for the better and neither has computer equity (cf. Grossman, 1998; Wise, 1997). Preliminary studies of advertising on the Internet show the same patterns being reproduced there (Kramer and Knupfer, 1997; Knupfer, 1997). The problem is not a lack of awareness on the part of advertisers but the power relations of the computer culture. Unless the technological frame enveloping computers

---

[14] Significantly, *Home Office Computing* was edited by a woman.

can be destabilized and reopened, attempts to create equity in computing for women and people of colour are unlikely to be successful.

Identity and Technology

The two case studies we have looked at in this chapter reveal much about the gendering of technology. Essentialist arguments (in the form discussed here) look at the ability of the cultural aspect to shape technology practice. Science and technology are associated with male values, language, and modes of thought which serve to exclude most women. Those who argue for the appropriation of technology look at the organizational aspect of technology practice. They say men, as a relevant social group, have used their power to exclude women and people of colour. Perhaps we can learn from both.

Constructionists are sometimes remiss in underemphasizing cultural factors. Gender roles are deeply imbedded in all aspects of society, including technology practice. They are part of an ongoing history which is more than just a "context"—they are part of all of us. On the other hand, gender itself is constructed. While the essentialists we have looked at here acknowledge this, for many the construction of gender-based value sets seems to have occurred some time in the past and are the culmination of a long philosophical and theological tradition, but they pay insufficient attention to the ongoing work of construction and reproduction.

The point is, both cultural aspects (values) and organizational aspects (power relationships) are reflexive. Neither can be understood without reference to the other. Computers are masculine machines today because they have been constructed that way. White, affluent men, as a social group, have appropriated the practice of computing. Their appropriation was achieved, in large part, through the use of socialization techniques and gatekeeping devices which defined computers in terms favourable to themselves and excluded most women and people of colour. At the same time, their task was made easier by inherited gender roles and gender-based value sets, language, and modes of thinking.

Because humans are technology-using animals, it is probably impossible to fully separate technology from our identities, nor is it, perhaps, desirable that we do so. Justice, however, requires a much greater concern with equity than we have seen in the computer culture so far. If we are to achieve a greater measure of equality in computing, we need to deconstruct the gendering of computers both through epistemological pluralism as a program and through specific measures to counteract exclusionary socialization and gatekeeping devices. We will need to confront *both* culture *and* power relationships.

Chapter Four

Venerating the Black Box

> Any sufficiently advanced technology is indistinguishable from magic.
>
> Arthur C. Clarke
> *Profiles of the Future*

> Magic is a flower that grows only in impotence.
>
> Rubem Alves
> *Tomorrow's Child*

Faust conjured spirits to give himself knowledge and power. In our century, we press buttons to do the same. The language of magic appeared to be a staple of popular discourse on computer technology in the 1980s. Computer products, advertising, and company names frequently used magic images or words. Magic language was particularly noticeable in media reporting on computers. As we move into the so-called information society, four hundred years after the scientific revolution began and two hundred years after the industrial revolution, why does the mass media's discussion of computers and related technologies use language such as "magic" and "wizardry"? Is it just hyperbole and advertising hype or, for all our pride in science and technology, do the wonders and terrors of sorcery still haunt our souls?

To answer these questions it will be necessary to examine the discourse surrounding the introduction of personal computers. The introduction of a major new technology, like any other form of social change, is potentially threatening to many people. Many ask themselves what do these changes mean and how will they affect me? One function of the mass media is to provide answers for these questions. The media play a crucial role in shaping perceptions, creating meaning, and directing the public's response to technological change. But the media are not neutral conduits of information. They serve both their own institutional interests and the interests of the social groups which control them. So when we find the language of magic in media discussion of computers and related technologies we have to ask how it creates meaning and why it is there.

In this chapter I will first describe technology and magic. Then I will analyze the contents of all the reports on computers and related technologies in *Time* magazine for a ten-year period, beginning with the advent of regular coverage of computers in 1979 (this period covers the crucial early years of the "personal computer revolution," a time when the meaning of the new technology was not yet established, through the stabilization of discourse in the mid- to late-1980s). I conclude by reflecting on some of the implications of this discourse.

Technology and Magic

Magical discourse seems to be alive and well in industrialized North America. During the ten-year period beginning in 1979, 36 percent of the 175 articles on computers in *Time* magazine used explicitly magical or religious language to refer to computers or to those who make, program, or use them (see Table 4.1). For example, the cover of *Time* for 16 April 1984 proclaimed "Computer Software: The Magic Inside The Machine." Words like magic, wizard, and wizardry appeared throughout the decade. Computers "conjure up" programs (*Time,* 31 January 1983: 65) while "teenage sorcerers" (*Time*, 3 May 1982: 54) used them for pranks or to practice the "secret arts of the computer age" (*Time*, 29 July 1985: 59). Sometimes the language was openly religious. Children "sit in communion" with their machines (*Time*, 21 September 1981: 60), while their elders, for whom "portable computers were gospel" (*Time*, 30 May 1983: 70), "like many other computer converts" recalled their "microelectronic baptism" (*Time*, 9 December 1985: 98). They struggled on "the long road to high tech heaven" (*Time* 3 January 1983: 17) while awaiting the "Second Coming for Steven Jobs" (*Maclean's*, 25 January 1988: 24). Even when explicitly magical or religious language was not used, sometimes what was described was implicitly magical, as when *Newsweek* proclaimed: "The pace of development is roughly akin to going from the Wright Brother's first airplane to the space shuttle in a decade" (30 June 1980: 51). Of course, such language did not dominate media coverage of computers, even in most of the stories in which it appeared. Magic was not the only trope used in media discussions of technology.[1] But the question which should be asked is: why was it there at all? Why, in an industrialized, secular society should some of the most advanced technology be described with the images and metaphors of magic? To

1 Dorothy Nelkin (1987) documents many of the trends and images in reporting on science and technology. In addition to magic, technology was described in images of the "frontier," the "cutting edge of history" and "breakthroughs" which will decisively reshape the future. Military metaphors were common, as technological mastery was portrayed as a "struggle" or "battle." Both the benefits and risks of technology were often described in apocalyptic terms. See also Corn, 1986.

answer this question we will have to examine the role of magical and religious language in discourse about technology.

As we saw in previous chapters, the meaning of any technology is established by its technological frame. Within the context of a technological frame, relevant social groups contend to impose their definition and understanding. Each of these groups will follow one or more strategies in this conflict. Should one of these groups be successful, we can say they have appropriated that technology, and it is quite likely that stabilization and closure will occur, which is to say, the victorious social group will impose its definition as the "universal" meaning of the technology and define its practices as "normal." After this happens the debate will "disappear" and the prevalent view will appear "natural" and a "consensus." Thus the development of any technology occurs in the context of a struggle over that technology's essential definition and its appropriation. The winners of the struggle define the technical culture, and become the experts upon which everyone else is dependent.

Control or manipulation of the media is one strategy by which a social group may strive to define and appropriate a technology. The mass media are powerful ideological resources (Gitlin, 1980; Lee and Soloman, 1991; Parenti, 1993). They provide a means by which contending social groups can "lengthen their networks" (Latour, 1987) by recruiting new allies and appropriating further resources.

The ideological power of the media lies in their ability to perform two functions. The first is ontological—they help to create reality. Dorothy Nelkin explains:

> For most people the reality of science is what they read in the press. They understand science less through direct experience or past education than through the filter of journalistic language and imagery. With the exception of an occasional television or radio notice, newspapers and popular magazines are their only contact with what is going on in rapidly changing scientific and technical fields, and their major source of information about the implications of such developments. (1987: 2)

Of course, not all people are influenced by the media in the same way. As Nelkin goes on to report, for issues about which people have little experience or prior knowledge the media "in effect defines the reality of the situation for them" (1987: 77). But in areas where individuals have experience or other sources of information the media tends to either confirm pre-existing biases or have its influence tempered by other sources of information.

The second function of the media is to establish or to undermine legitimacy and authority (Meyrowitz, 1985). By chosing who and what to cover and by adopting a particular language the media bestow facticity on the relations of power so as to augment or undercut authority. Generally those involved with science and technology have had their legitimacy bolstered by the media. While the context and background of individuals are as important in conferring legitimacy as in defining reality, research shows that "the media portrayal of science and technology as esoteric and arcane yet a source of authority and broad-ranging expertise" is widely accepted by the public (Nelkin, 1987: 76).

Whether defining reality or bestowing legitimacy, the media must maintain plausibility. An account which lacks plausibility for too many people will quickly lose credibility and may actually delegitimate its source. One way journalists protect their own credibility is through professional norms of "objectivity" (Tuchman, 1978); another is to reflect what they perceive to be their audience's preferences, concerns, and values (Nelkin, 1987). Reporting is thus a reflexive activity. The media both create and mirror reality. They try to "objectively" describe an actuality which they themselves have helped to construct.

The agency of this reflexivity are *media frames*, which Todd Gitlin defined as "persistent patterns of cognition, interpretation, and presentation, of selection, emphasis, and exclusion, by which symbol-handlers routinely organize discourse, whether verbal or visual" (1980: 7). A frame acts as a filter for reporters and editors. "This frame organizes the world for journalists," says Nelkin, "helping them to process large amounts of information, to select what is news, and to present it in an efficient form. Their metaphors, descriptive devices, and catch phrases are expressions of this frame" (1987: 9). A frame is also a focus for how the news is presented. The same metaphors which helped the reporter organize information become the means through which reality is created for the audience. Particularly potent images or metaphors, once part of a media frame, can go on reciprocally shaping the social world and the media accounts of that world for a long time.[2]

Although control or manipulation of the media may be an effective strategy for those contending to define and appropriate a technology, not all relevant social groups are equally able to use it: the ability to "make news" is an aspect of power (Logan, 1977). Those who lack the power to access the nodes in the news-gathering net (Tuchman, 1978) are much less likely to influence the shaping of media frames, or indeed, to be covered at all.[3] This is particularly true in reporting on science and technology.

---

[2] See Weart (1988) for an outstanding analysis of this in the field of nuclear technology.

[3] There is, of course, a reciprocal relationship between access to the news-gathering net and power. Those with power (e.g., the mayor, the President) have the ability, by virtue of their position, to command media attention. The nodes in the net are typically the "news beats" located in the offices of the powerful (e.g., city hall, the White House). On the other hand, those who are able to gain media attention also gain power (e.g., Greenpeace).

Science reporters are especially dependent upon scientists, engineers, and the organizations for which they work as sources—much more so than, say, political reporters who have a wider variety of "informed opinion" to draw upon (Nelkin, 1987). With the exception of "media-wise" groups such as Greenpeace, media frames of technology are usually shaped by government, industry, and the large research institutions.

At this point technological frames and media frames begin to merge. The scientists' and technicians' "theories, goals, and tacit knowledge"—and particularly their images and metaphors—become part of the reporters' "pattern of cognition, interpretation and presentation." As the frames merge, media frames magnify technological frames. The rhetoric of those social groups that serve as sources (scientists, engineers, corporate and government public relations people) is amplified (and filtered and modified) through journalists to other social groups that make up their audience. This expansion of discourse enlarges technological frames by bringing more groups into the discussion. A media campaign is thus often an effective strategy for enlisting allies (especially, in the case of technology, potential customers). A successful campaign can sway "public opinion," and thus also affect governments. Most significantly, powerful social groups can use the media to "set the agenda," that is, to focus public discourse on issues those groups want discussed, framed as they want them seen. To use an example from Weart (1988), the nuclear industry wants the agenda to be the increasing need for energy while environmentalists want to frame the debate in terms of safety and pollution.

Analysis of media frames gives access to a fragment of a technological frame in the process of amplification, modification, and diffusion to a larger number of relevant social groups. Given the close relationship between reporters in this area and their sources, we should expect that stabilization and closure of the technological frame will be reflected in media coverage.

This is the context in which we will examine the structure of *Time*'s reporting on computers and related technologies. As the continent's largest-circulation newsmagazine, *Time*, together with a very few other "flagship" publications, plays a particularly powerful ideological role, and is second only to television in constructing social reality (Gitlin, 1980, Tuchman, 1978). Moreover, in the case of science and technology, the print media may be more important than television (Nelkin, 1987). Although *Time* represents only a fragment of the total discourse on computers, it is certainly a significant one. I therefore expect to find that the

---

For those without institutionalized power, however, media access is always precarious and dependent upon actions which are "newsworthy," which in practice usually means more dramatic or sensational than the last time the media noticed them (Gitlin, 1980). The President is newsworthy every day, Greenpeace is not.

use of magical language in *Time*'s coverage will in effect favour the interests of one or more powerful social groups even as it attempts to define the meaning of the new technology for the wider public.

In order to understand *Time*'s use of magical language, I examined its reporting from two perspectives. First, I attempted to estimate the extent to which magical language was used by performing a content analysis and counting all the occurrences of explicitly magical and religious language in *Time* reports on computers and related technologies between January 1979 and December 1988 (as mentioned at the beginning of the chapter, this period covers the crucial early years of the "personal computer revolution" through the stabilization of discourse in the mid- to late-1980s). A word was considered explicitly magical or religious if that is its ordinary, everyday meaning. For example, words like *magic*, *wizard*, *sorcerer*, and *conjured* are explicitly magical whereas *gospel*, *communion*, *baptism*, and *heaven* are explicitly religious. Words which were not explicitly magical or religious in their everyday meaning (e.g., wondrous) were not counted. The emerging patterns reveal the structure of *Time*'s coverage.[4] Second, this discourse was qualitatively examined through a detailed anthropological profile of magic to which media discourse was compared. The aim of this hermeneutical interpretation (Geertz, 1973; Ricoeur, 1976), was to discern how the use of images and metaphors helped to create meaning. Together these two approaches should help us to understand how magical and religious language was used by *Time* to define the new technologies.

Magic and Religion

A trichotomy between science, magic, and religion was a significant issue for the founders of the sociology of religion, such as Emile Durkheim, Max Weber, and Bronislaw Malinowski. According to Durkheim, magic and religion are closely related. He defined religion as "a unified system of beliefs and practices relative to sacred things, that is to say, things set apart and forbidden…which unite into one single moral community called a church, all those who adhere to them" (1915: 62). Magic is also a system of beliefs and rituals pertaining to sacred things, but in contrast with religion, "there is no church of magic" (1915: 60). The magician serves clients, not a congregation or a moral community.

---

[4] As a control I checked *Time*'s coverage against selected cover stories and articles in *Newsweek* and *Maclean's* (Canada's major newsmagazine) for the period under study. A random sample of their reporting on computers and related technologies for the same period revealed only a few significant differences. *Newsweek* appeared to use less explicitly magical language but to be even more utopian. *Maclean's* coverage was much weaker and more slanted towards business.

Although Weber discussed magic and religion extensively, he never presented a clear definition of either term. The key element in both was the possession of charisma. For Weber, charismatic leadership was the basis for one type of legitimate authority (1947: 358-63). He defined charisma as:

> A certain quality of an individual personality by virtue of which he is set apart from ordinary men and treated as endowed with supernatural, superhuman, or at least specifically exceptional powers and qualities. These are such as are not accessible to the ordinary person, but are regarded as of divine origin or as exemplary, and on the basis of them the individual is treated as a leader. (1947: 358-59)

To Weber, "the magician is the person who is permanently endowed with charisma" (1922/1963: 3) but like Durkheim, he distinguishes between priests who were functionaries of a cult and "magicians, who are self-employed" (1922/1963: 29). If the line between religion and magic was blurred in Weber's work, the distinction between them and science was not clear either. Both magic and religion are oriented to "this world," exhibit "relatively rational behaviour," and their ends are "predominantly economic" (1922/1963: 1). For Weber, then, the possession of charisma, some form of extraordinary power, was the defining characteristic of both magic and religion, but the boundaries were indistinct.

Bronislaw Malinowski (1925/1948) also described a science-magic-religion trichotomy. The essence of magic is *mana* (like Weber's charisma, a manifestation of extraordinary power). This distinguishes it from science. At the same time, magic shares with science an orientation towards achieving some concrete goal or purpose. To Malinowski, this is the difference between religion and magic. "In the magical act the underlying idea and aim is always clear, straightforward, and definite, in the religious ceremony there is no purpose directed towards a subsequent event" (1925/1948: 38). Magic relies upon extraordinary power, upon mana, but is oriented solely to achieving practical, "this worldly" ends.

Today's definitions generally either follow one or the other of these "classic" definitions (Stark and Bainbridge, 1985) or attempt a synthesis of them (O'Keefe, 1982). Perhaps one of the best short definitions is offered by Joachim Wach: "Magic means to force the numen to grant what is desired" (1944: 353). Most take the view that magic is either a pseudo-science or a deficient form of religion. For those with an evolutionary viewpoint (de Bolt, 1969), magic represents a "primitive" form of belief which was replaced by "modern" and certainly non-magical ways of thinking. Evaluations differ. Some hold that "magic must rank among the greatest of men's delusions" (Webster, 1948: 506) while others contend that early science was a continuation of the quests and methods of alchemists and magi (cf. Eliade, 1978; Merchant, 1980; Seligmann, 1948). Thus, according to most definitions, the difference between magic and religion is more a matter of institutional differentiation than of substance or function. Both address the numinous: religion within the context of the cult, magic outside of it. In this study, however, we are discussing discourse in the media rather than cultic practices. Institutionally grounded distinctions between magic and religion lose their significance. Instead, this phenomenon is an implicit religion. While differences in vocabulary may be noticeable, what is significant for this study are appeals to the numinous. Expressions such as "the magic inside the machine" or "microelectronic baptism" may have different histories, but both speak of technology in the language of the numinous. For the purposes of this study, all appeals to the numinous will be equivalent.

Frame Analysis

Time began regular reporting on computers in 1979, two years after the first microcomputers became commercially available. Reporting was institutionalizied when a "Computers" department was created in 1982.[5] This department was discontinued in mid-1987 and replaced by a more general one called "Technology." From only three in 1979, the number of articles on computers increased dramatically and reached a peak in 1983, then trailed off (see Table 4.1; Figure 4.1). A better measure of the extent of coverage is the number of column centimeters devoted to a story. The total of column centimeters that *Time* dedicated to computers increased by an average of 195.25 percent per year between 1979 and 1983, dropped sharply in 1984, and then, with a few fluctuations, drifted downwards.

The intensity of coverage of a story can be measured by the average length of the articles allotted to it. Articles on computers averaged 65.4 column centimeters over the ten years studied, with a significantly higher average in 1979, 1980, 1982, and 1983. One measure of the prominence a magazine gives a story is the number of covers devoted to it. Seven of the nine cover stories *Time* did on computers in this period were done between 1980 and 1984, including the "Machine of the Year" cover in January 1983. These measures of the extent, intensity, and prominence of stories about computers suggest a period of growing awareness from 1979 to 1981 and a period of peak enthusiasm from January 1982 until April 1984, followed by a period of stabilization (see Table 4.2).[6]

---

[5] For the importance of this kind of institutionalization in the news-gathering net, see Tuchman, 1978.

[6] These periods are an artifact of the discourse itself, and are not connected to changes in personnel. An analysis of identified writers and reporters reveals no pattern associated with any particular individual. People came and went over the ten-year span, but their

Table 4.1

***Time* Magazine Discourse on Computers**

YEAR	NUMBER OF ARTICLES	COVER STORIES	COLUMN CENTIMETERS Total	Average	ARTICLES USING MAGICAL OR RELIGIOUS LANGUAGE Number	Percent
1979	3	-	219	73	1	33.30%
1980	6	1	539	89.83	3	50.00%
1981	12	-	602.5	50.2	6	50.00%
1982	17	3	1607.5	94.56	7	41.18%
1983	32	2	2526.5	78.95	14	43.75%
1984	27	1	1637	60.63	11	40.74%
1985	24	-	1231	51.29	7	29.20%
1986	22	1	1317	59.86	6	27.73%
1987	14	-	830.5	59.32	3	21.40%
1988	18	1	934.5	51.92	5	27.78%
TOTAL	175	9	11444.5	65.4	63	36.00%

Figure 4.1
***Time* Magazine Articles on Computers**

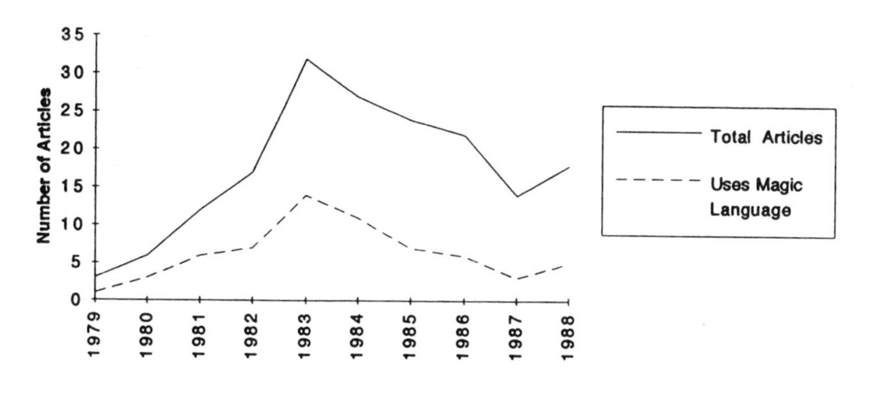

comings and goings are not associated with any change in the style or content of the discourse. This applies to the use of magical and religious language as well. The same individuals would use magic language in some stories but not in others. We should note that until 1985 all the identified writers on computers (as opposed to reporters) were men.

Table 4.2

Periods in *Time* Magazine's Reporting on Computers

	1979-1981	1982-April 1984	May 1984-1988
Duration			
Length (in months)	36	28	56
Percent of total	30%	23%	47%
Total number of articles			
Number	21	58	96
Percent of total	12%	33.14%	54.86%
Total column centimeters			
Number	1360.5	4905	5199
Percent of total	11.89%	42.86%	45.43%
Average column centimeters	64.79	84.57	54.16
Cover stories	1	6	2

Time's coverage during the period between January 1982 and April 1984 was particularly intensive.[7] Fifty-eight of the 175 articles on computers, or 33.14 percent of the total, appeared during those twenty-eight months. In total, articles on computers during this period occupied 4,905 column centimeters, or 42.86 percent of the ten-year total. The average length of 84.57 column centimeters per story was well above the 65.4 column centimeter average for the ten-year total. Six of the nine covers devoted to computers during the ten years under study appeared during these twenty-eight months. After April 1984, the number of articles declined slowly, but the average length of stories dropped sharply and computers were never again given such prominence.

The magical and religious language used by *Time* in its stories about computers is part of this broader pattern. The numbers of articles in which explicitly magical or religious language appears parallels the total number of articles on computers, rising sharply to a peak in 1983 and then trailing off (see Figure 4.1). Explicit magic language was found in 46 percent of all articles on computers from 1980 to April 1984, and then dropped sharply to an average of 28 percent of the articles between May 1984 and 1988.[8] The amount of magical language found within these

---

7 The "signposts" of this period are the cover stories on 18 January 1982 "Gronk! Flash! Zap! Video Games Are Blitzing the World" and 16 April 1984 "Computer Software The Magic Inside the Machine." One reviewer noted that the beginning of this period shortly follows the introduction of the notoriously "user unfriendly" IBM PC while its conclusion roughly corresponds to the introduction of the much more "user friendly" Macintosh from Apple. While it is fascinating to speculate that this is more than coincidence, I can find nothing in the texts themselves to justify that conclusion.

8 The break in 1984 is quite pronounced. From January to April 1984 magic language appeared in 55.56 percent of stories on computers, from May to December it is found in only 33.33 percent of such articles. As was mentioned above, this break is not associated with any change in personnel.

stories also varied. Sentences containing explicitly magical or religious language occurred on average 1.8 times per story in 1979-1981, rose to 2.92 per story between January 1982 and April 1984, and after that declined to an average of 1.7 per story. Thus, the use of explicitly magical and religious language was widespread but not intense and significantly more common during the period of peak enthusiasm than later.

There were also structural changes in how explicitly magical and religious language was used. If we think of a news story as a frame, it has both a "center" and a "rim." The rim is what "frames" the story, bracketing the central text, placing it in perspective, and marking its boundaries with the surrounding context (Brown and Yule, 1983; Gitlin, 1980; Tuchman, 1978). Words or images which appear on the rim connect the substance of the story with an assumed context; providing continuity with previously established meanings and a "flag" which alerts the reader to something new or different (the center of the story). In a magazine story, the rim usually consists of the headlines, subheads, and first and last paragraphs. When we examine sentences in *Time*'s reporting on computers in which explicitly magical or religious language occurs, the ratio between the number appearing in the center of the story and on the rim varies from one period to the next. During the first period, 1979-1981, 39 percent of magical language is found on the rim of the story. In the second period, January 1982 to April 1984, the percent on the rim drops to 25 percent while in the final period it rises to 52.17 percent.

All this indicates a changing function for magical language. During the first period it signals change and novelty. During the second period, the more frequent use of explicit magical language and its predominant location in the center of the frame indicates that it plays a role in defining and giving meaning to the new technology. After April 1984, there is a relative scarcity of explicit magical language and it is more frequently located on the rim of the frame. The position on the rim apparently now has a different function. Personal computers are no longer "new;" their meaning has been established and now it is only necessary to make reference to previously accepted definitions. This implies that discourse has stabilized.

Structural analysis of *Time*'s reporting points out the small but not insignificant part which explicitly magical and religious language played in the frame, but it misses the much larger, but implicit, use of such language in defining the new technologies.

A Profile of Magic

A useful profile of magic has been developed by anthropologists Rosalie and Murray Wax (1962). They describe an encompassing magical worldview in which beliefs and practices meld. According to their characterization, the magical worldview differs from the modern in four ways: its relationships, its understanding of power, of logic, and the role of the person. In each of these areas, what Wax and Wax describe as magic bears intriguing similarities to the media's discussion of computers.

The first quality of the magical worldview is the relationships within it. Wax and Wax cite H. and H. A. Frankfort, "For modern scientific man the phenomenal world is primarily an 'It'; for ancient—and also for primitive—man it is a 'Thou'" (1962: 180). Humans and natural entities are bound together both socially and emotionally because "the intimate linkage between man and non-man is manifested in the fact that the former is not intrinsically distinguished from the latter in physical makeup" (1962: 182). The magical world is one where humans and animals engage in all forms of social interaction, where plants speak and stones weep. In the industrialized world, we lack this sense of intimacy between humans and the non-human world (Merchant, 1980). To us, the world is mere stuff. When we start speaking of nature as "overburden," the strip mines are not far behind. But computers are frequently spoken of differently. Perhaps because computers sometimes seem to be made in our mind's image, people react to them differently than to other machines. As Sherry Turkle found, children are often unsure whether computers are alive or not (1984a: 29-63). *Time* magazine is not completely sure either.

Anthropomorphism is endemic to discourse about computers. Much of this comes from the industry's jargon. Computers are said to be "smart" machines with "memories," who "talk" to each other in "languages" and develop "artificial intelligence." This kind of anthropomorphism was common throughout the 1980s. If computer professionals speak this way, we can expect that the popular media will do it as well. But the newsmagazines went well beyond this. *Newsweek*'s cover boldly proclaimed "Machines That Think" (30 June 1980). *Time* described computer toys as "clever new playmates" and "as teachers they can form bonds of a sort—friendships?—with their pupils" (10 Dec. 1979: 68). One computer "plays imaginative court to details" (*Time,* 31 Jan. 1983: 65) while others are "expert assistants" (*Time,* 2 Sept. 1985: 43). The machines were bestowed with emotions. The article on computer toys had them pondering, cheerful, and "becoming impatient within a couple of minutes when its opponent is thinking" (10 Dec. 1979: 71). Teaching programs were said to be "very patient and non-judgmental" as well as polite. (*Time,* 21 Sept. 1981: 60). Reportedly, people responded to them just as emotionally. *Time* worried about "computer addicts" who "become obsessed with the machine, at least in the eyes of other family members" (15 Oct. 1984: 103). Indeed, *Time* added, "the most successful machines have a built-in emotional component, something that connects the tools in the computer with the whims of its user" (24 Oct. 1988: 76). In these human-machine relationships, the machine was frequently portrayed as the active partner. Computers were spoken of in the active voice, as if they had volition. In the military, microchips "perform such crucial tasks as aiming guns and navigating flights" (22 Oct. 1984: 60) reported *Time*, and "computers have taken charge of the logistics of the Olympic games" (9 July 1984: 52). The machines also got sick. "Computer virus" is an industry term but once again *Time* went well beyond metaphor. According to *Time*, a computer virus is "a self-replicating organism" (26 Sept. 1988: 54) which is "contagious" and may cause an "epidemic." Help was near, however. A "vaccine program" may "protect an infected disk," unless the machines "suffer a relapse" (26 Sept. 1988: 55). All this kind of talk could lead one to ask—paraphrasing a Native American quoted by Wax and Wax (1962: 182)—"ain't computers people?"

The second quality of the magical worldview is Power. Wax and Wax use the word (capitalized) to describe the dynamics of the magical world. Arising out of the context of the relationships between humans and nature, Power is generated through rituals and, like electricity, may be accumulated and then employed or discharged. People with the modern worldview usually believe Power to be miraculous or supernatural, something involving the suspension of natural laws. This is not the case with the magical worldview. According to Wax and Wax:

> Viewed by a person in the magical world, Power is awesome and wonderful but, at the same time, it is an intrinsic feature of the natural order, manifesting itself in much of what we (but not he) would consider "common" or "ordinary." Thus any activity of note or worth can only occur because of it. (1962: 182-83)

Magical Power is neither mechanical nor permanent. People learned the rituals of Power from magical beings or ancestors. If they lose favour, or grow careless or arrogant, they may just as easily lose it.

The quest for Power is the great Faustian theme of modern civilization. Computers are very much a part of that quest. "Power" and "powerful" are among the words most commonly associated with them. Here again, media discourse went beyond industry jargon or advertising hype to an expression of Power as *mana*. Typical is a report of a fourteen-year-old boy as saying: "I love these machines. I've got all this power at my fingertips. Without computers, I don't know what I'd be. With them, I'm somebody." (*Time*, 3 May 1982: 49). The computer is a talisman or focus through which all limits could be transcended. "In the hands of a professor who really believes," says a student at Drexel College, "it seems the computer can do miracles." (*Time*, 21 Oct. 1985: 83). This Faustian quest to transcend limits may be trivial or world-shaking. An example of the former was *Time*'s celebration of a new folk hero. A fifteen-year-old named Steve Juraszek achieved "what is very close to being impossible" (18 Jan. 1982: 49), a sixteen-hour "enchanted run" (18 Jan. 1982: 52) on the video game *Defender* which scored a world record 15,963,100 points. But quests for Power may be more serious. A story on computer hackers reports on "the unspoken assumption among crack computer programmers and engineers that they could straighten out the world by dint of their intelligence if they could only get their hands on the control box" (*Time*, 3 Dec. 1984: 71).

When *Time* discussed "this transformation of the young" (3 Jan. 1983: 16) into the Agents of History, they did so for a specific reason. Young people know the rituals which put them in touch with Power. It is called programming. Computers were mysterious things, full of "names that are as mystifying to an outsider as the secret password of an esoteric cult" (*Time*, 3 Jan. 1983: 27). *Time* was quite willing to put its readers "in the know" by introducing them to a little of the lore in the form of glossaries and even scraps of incantations:

```
EMIS-HIT?      LDA JETY
               SBC EMIST
               CMP #10
               BGE EMISEXIT
PLAYR-HIT?     LDA EMISEXIT
               STA JETCOND
```

(16 April 1984: 74)

But with these tantalizing glimpses into the realm of Power came warnings that the rituals were only for the initiated. Those who would use magic must pay a price. "A programmer can frequently spend 18 hours a day at a terminal working on a difficult problem. That fanaticism allows very little time for ordinary human pursuits; programmers often wryly characterize themselves as 'computer nerds'" (16 April 1984: 73). Adepts—we call them hackers—were a people set apart. We should note that this kind of language parallels the use of explicit magic language and after April 1984 became uncommon in the newsmagazines' coverage of computers.

Perhaps the place Power was most manifest in *Time* was the business pages. Business success or failure was portrayed as personal, inexplicable, and dependent upon Power. Entrepreneurs were regularly pictured as archetypal heroes. Exploding profits were invariably cited as confirmation of their magic, failure is a sure sign that Power has deserted them. Thus in the heady days of the early 1980s, a period of rapid growth in the computer industry, *Time* said of IBM, "the firm has always had a special mystique" (24 Aug. 1981: 44). Upon introduction of its first personal computer, a *Time* source claimed "some people were convinced that IBM would be unveiling a new Holy Grail" (24 Aug. 1981: 44). But when an industry shake-out occurred in the mid-1980s, the Power seemed to be gone. "The IBM label has lost some of its magic" one analyst was reported as saying (*Time*, 21 July 1986: 45). Profitability returned to the industry in 1987, but magic language did not return to *Time*'s business pages.

Underlying Power in the magical worldview is a unique form of logic. According to Wax and Wax, the world of magic differs from our own in the very basis of understanding and explanation. Our world of genetics and quantum mechanics rests upon chance. We accept the logic of contingency. The magical world does not. Everything is controlled and explained through the causal dynamics of Power.

> The magical world has a rigorous causal scheme of a pragmatic and retroactive character: Success demonstrates the presence of Power, failure its absence. Life is not an accidental succession of chance occurrences, but exhibits the presence of varying kinds and extents of Power as affected by relationships among beings. (1962: 184)

In the magical world, illness, for instance, is not the result of a chance meeting with a microbe but the result of a malignant will. Perhaps the protective rituals were done improperly. Perhaps the victim is bewitched or an an evil spirit has proven even more powerful. Recovery means determining the actual (magical) cause and performing the correct propitiatory or cleansing ritual. "Within the magical world," say Wax and Wax, "ritual is the focal point of existence" (1962: 184). The aim may be to manipulate the world or it may be protective in nature, as O'Keefe also emphasizes (1982: 262ff.). But in whatever form it may take, the aim of ritual is to put the person in contact with Power.

In most of its reporting, *Time* assumed the normal contingencies of modern life. But in those stories which discuss the ebb and flow of Power a different logic comes into play. Events assume a great sense of inevitability and Power did not operate without cause:

> Whatever its variations, there is an inevitability about the computerization of America. Commercial efficiency requires it, Big Government requires it, modern life requires it, and so it is coming to pass. But the essential element in this sense of inevitability is the way in which the young take to computers: not as just another obligation imposed by adult society but as a game, a pleasure, a tool, a system that fits naturally into their lives. Unlike anyone over 40, these children have grown up with TV screens; the computer is a screen that responds to them, hooked to a machine that can respond the way they want it to. That is power. (*Time*, 3 Jan. 1983: 15)

With such a strong sense of inevitability, it is perhaps not surprising that discussion of future prospects was not scientific prediction or forecasting to be tested but prognostication based upon faith. When the prophecy failed, the logic of inevitability was applied retroactively. Thus when the companies which were touted as "up and coming" success stories in 1981-83 failed in 1984-85, *Time* took the loss of Power in stride by talking of the lessons of history and "the inevitable changes that come with rapid growth" (30 Sept. 1985: 65). The logic of magic is unfalsifiable.

The fourth quality Wax and Wax see as characteristic of the magical worldview is the special place of the person. The magical world has no place for individualism. Myths and legends are populated by heroes, but they lack both individuality and character development as measured by modern standards. The heroes of magic are defined by the presence or acquisition of Power. "Many of the tales are success stories," say Wax and Wax, "beginning with a young person whom folk consider a n'er-do-well and ending with his demonstration of Power" (1962: 185). Character or morality are irrelevant, only the presence or absence of Power matters. The actual magic world is thus populated by the likes of Coyote the Trickster or the Monkey King rather than Frodo the Hobbit.

In the pages of the newsmagazines, the people who make and use computers did not quite fit the profile. We were still presented with types—*Time* gave us mythologized heroes and villains—but the pattern of the myth was more modern than what Wax and Wax describe for the magical worldview. Still the magical language around these people was stronger and persisted longer than in other stories in the magazine.

The people whose portrayal in *Time* came closest to matching that of the profile were the young "whiz kids" and hackers. As we have already seen, *Time* gave young computer users a very special place. They became the Sorcerer's Apprentice. In only a few cases did *Time* develop the personality or biography of the subject of the story, that is, make them an "individual." In all cases what makes them newsworthy—what made them a type—is the possession of Power. Computer virtuosos were invariably portrayed as young. School children were shown as having a natural affinity for computing not shared by adults. As possessors of Power, they inverted traditional roles, "tasting the heady pleasure of teaching their own teachers" and "instructing their parents as well" (*Time*, 3 May 1982: 53). Hackers were a breed apart. By their dress and demeanour they were depicted as perpetual adolescents, no matter what their age (*Time*, 3 Dec. 1984: 71-72). As a type, the Sorcerer's Apprentice was a figure of both promise and menace—precisely the amorality discussed by Wax and Wax. They may use their Power for good, as did Will Harvey, aged sixteen, who "sat down with his Apple and an introductory music text and came up with a program that is making even professional musicians stand up and shout 'Bravo'" (*Time*, 17 Oct. 1983: 49). Or even more frequently, their Power may be devoted to pranks, crime, or the creation of viruses.[9] What made them significant was the Power itself.

The other group to whom *Time* accorded mythic status were the Entrepreneurs. There was a clear difference here among those who make computers and software. Scientists and engineers were routinely referred to as "magicians" and "wizards" right through the decade, but otherwise relatively little attention was paid to them. It was businessmen (and all of them were men) that *Time* celebrated. Some of these were inventors, programmers, or engineers as well, but it was their prowess in the marketplace that elevated them. While *Time* tried very hard to find some element of Horatio Alger in nearly every executive (cf. 3 Jan. 1983: 20-21), the person who most fully achieved heroic status was Steven Jobs.

The successes and failures of Jobs was a saga spread out over ten years. Early stories emphasized Jobs's youth, "humble" origins (a college dropout and "self-made engineer"), unusual lifestyle ("a confirmed fructarian"), and amazing financial success (*Time,* 5 Nov. 1979: 65). By 1983, *Time* had grown rapturous over his "fairy-tale success" (3 Jan. 1983: 17). While recognizing that Steve Wozniak was "the true technological wizard" (3 Jan. 1983: 17) behind the Apple computer, *Time* portrayed Jobs as a religious visionary. Possessing a "smooth sales pitch and a blind faith that would have been the envy of the early Christian martyrs" (3 Jan. 1983: 17) he "is positively hypnotic when he takes the computer gospel to the young" (3 Jan. 1983: 19). Steven Jobs had Power. His success (as measured by his financial wealth and the price of Apple stock) were proof of it, overriding personality quirks. But the magic was not to last. Jobs lost the Power, his position, and the admiration of *Time*. Those personality traits which were colourful when he was successful now loomed large as the reasons for his failure. *Time* dismissed him with the words of a venture capitalist: "It's good news for Apple that he's out of their hair" (30 Sept. 1985: 65). Bereft of Power, Jobs dropped from the magazine pages for a while. When he resurfaced, his new company NeXT was seen as an attempt "to redeem himself." Jobs regained his Power. *Time* reported one industry editor as saying, "I arrived a non-believer, and I came away a

[9] Even when *Time* recognized—rather late in the decade—that most computer crime was not committed by teenagers, two of the four specific examples cited *did* involve young people (17 Feb. 1986: 49). This report did not seem seem to affect subsequent coverage of the issue.

convert" (24 Oct. 1988: 76). His new machine was "a computer with soul" (24 Oct. 1988: 77).[10] Jobs had returned, almost like a messiah.

The Entrepreneurs as a group, and Steven Jobs in particular, were endowed with more of an individual personality by *Time* than Wax and Wax claim to be characteristic of the magical world. However, a sense of magic is there, and would not be inconsistent with Weber's understanding of charisma. And in some respects, especially the belief that what *really* counts is Power, the people in *Time* do fit the profile. "If a person has obtained Power," claim Wax and Wax, "what he did was correct, and if he lacks Power, what he has been doing is incorrect" (1962: 185). This statement certainly applies to *Time*'s coverage of the Entrepreneurs.

Thus for each of the features of the magical worldview described by Wax and Wax we have found examples, not from ancient sagas or anthropologists' field notes, but from the pages of North America's largest circulation newsmagazine. Although roughly parallel to the patterns uncovered for the use of explicitly magical and religious language, the implicit use of magical themes is far more extensive and integral to *Time*'s discourse on computers and related technologies. Together the language and themes make up a part of computing's technological frame and play a role in creating and stabilizing the meaning of the new technologies. The remaining question is why?

Machines as Metaphors

Why did *Time* use magic language to talk about computers and related technologies? The answer is complex. There are many interrelated levels of meaning in *Time*'s discourse.

At the most superficial level, magical language provides a lively and appealing way to talk about technological change to a mass audience. *Time*, after all, has a long reputation for vivid writing. It sells magazines. By building readership among the middle- and high-income males who were particularly interested in the new technology, *Time* was better able to sell an audience to its advertisers. But this superficial reading does not explain the patterns we uncovered or why the use of magical language increased and declined.

If, as Lakoff and Johnson write, "the essence of metaphor is understanding and experiencing one kind of a thing in terms of another" (1980: 5), by using magic as a metaphor for technology and the people around it, *Time* was telling us that technology was like magic. In what ways were they similar?

---

[10] The other newsmagazines were even more ebullient. *Maclean's* headlined "Second Coming For Steven Jobs" (25 Jan. 1988: 24); *Newsweek* devoted a cover to his new machine (24 Oct. 1988).

Technological change may create uncertainty. Few know how the new machines function. Their effects are unknown, especially when they change the everyday routines of work (or threaten to replace workers altogether). Conventional machines with which people are familiar and comfortable are not discussed in magical terms (indeed, they are not newsworthy at all). Computers may have been magical, toasters were not. Compounding the uncertainty inherent in any technological change was the perception that computers were machines which could think—the very quality which has often been used to define humanity. *Time* said as much. In an essay entitled "The Mind in the Machine," *Time* observed that, "When machines possess artificial intelligence, like computers, the human fear of being overtaken seems both more urgent and more complex" (3 May 1982: 60). To *Time*, the mind was itself "an enchanted thing" (3 May 1982: 61), so a machine seen to be built in the image of the mind was doubly so. Computers were powerful, but also mysterious. Their power was ours to use but not to understand. In a real sense Arthur C. Clarke's famous "Third Law" is true: "Any sufficiently advanced technology is indistinguishable from magic" (1973: 21). When technology is a "black box," it becomes magical.

As a magical black box, computers were portrayed as a source of hope amidst fear. As the division of labour, and with it the information explosion, continue to intensify there is more and more knowledge which we not only cannot master but for which we must trust some "expert" (Adorno, 1974: 78). Extreme specialization means that we lack control over many aspects of our lives. Computers and related communications technologies intensify this situation while promising individuals a means of coping with it. *Time*'s reporting reflected this. Vague anxieties were given focus through the magic language surrounding the new technologies. For example, *Time* reported: "Many Americans concerned about the erosion of the schools put faith in the computer as a possible saviour of their children's education, at school and at home" (3 Jan. 1983: 15). At one level, this passage is yet another proclamation of a "technological fix." But in using language like "faith" and "saviour," *Time* is invoking the numinous. Why? As Rubem Alves reminds us, "Magic is a flower that grows only in impotence" (1972: 81). When people feel powerless, they turn to magic. "The function of magic is to realize man's optimism," wrote Malinowski, "to enhance his faith in the victory of hope over fear. Magic expresses the greater value for man of confidence over doubt, of steadfastness over vacillation, of optimism over pessimism" (1925/1948: 90). Magic is a declaration of hope in the face of fear and powerlessness. *Time* seems to be telling its readers to have faith, no matter what anxieties you may feel, be confident in the future. The invocation of the numinous elevates this above a mundane prediction of a quick fix to an article of belief. However, in *Time*'s usage, both hope and fear are translated in a way that reinforces the status quo.

This can be seen particularly clearly in the way in which *Time* discussed young people and computing. As we have already seen, *Time* reported that young people had a natural affinity for computers; indeed, it presented them as a type—the Sorcerer's Apprentice. In the most literal meaning of this metaphor, young people were real apprentices. They were understudies for the true magi, the Entrepreneurs. Implicit in the discussion of a young person's achievement was the possibility of commercial success. Students who were able to market their programs were given prominence (e.g., 2 March 1981: 43; 5 Oct. 1981: 83; 3 May 1982: 59). The youth of Steven Jobs and Bill Gates of Microsoft was held out as an example. Implied in these stories was the message that your child could do it too. The new technology placed success within reach.

At another level, young computer users were threatening. Like the Sorcerer's Apprentice, the menace of the teenage hacker was Power out of control. Like youthful street gangs, hackers represented the potential for random and meaningless destruction. (That most computer crime was committed by disgruntled employees rather than irresponsible teenagers was not a fact that *Time* liked to report). The message here was the need for control and discipline.

Finally, young computer users represented the hope of the next generation. It has been a central American myth since the Depression and Second World War that children will do better than their parents, making their parents' sacrifices worthwhile. *Time* was tacitly telling parents threatened with loss of self-image or employment that you may not be coping with technological change but don't worry, your children will do well. However uncertain things may appear now, the future was bright. In all of these accounts, *Time*'s reporting focused discourse through widely available typologies and previously constructed definitions. The effect was to stabilize meaning and thus diminish the tensions and anxieties brought about by technological change.

This stabilization of discourse was to the advantage of those who were best able to appropriate the new technologies. As we have seen, stabilization is itself a sign that one or more social groups have been successful in appropriating a new technology. *Time*'s contribution to this discourse was to promote definitions and understandings favourable to business. In creating the Entrepreneur as a type and endowing entrepreneurs with Power, *Time* was investing them with authority. This is why *Time* bestowed charisma upon Steven Jobs. If Weber's theories (1947: 360-62) of charismatic authority are correct, *Time* was tacitly saying that the corporate elite were leaders who ought to be obeyed. They had the Power to augur the mysteries of the market. Others who may have possessed magic were subtly marginalized. Scientists were routinely referred to as wizards but they never really escaped from the black box. They were wondrous, inexplicable—and marginal. Children were Entrepreneurs-in-waiting so long as they were "well behaved," and were portrayed as threats when they were not. Indeed, of the various possible threats created by computers, *Time* paid far more attention to the potential damage caused by hackers and their viruses (which could be damaging *to* business) than it did to technology-caused unemployment or the invasion of privacy (which were largely caused *by* business). Similarly, the claims of other relevant social groups were generally ignored. The extensive debate over women and computing which was going on in the mid-1980s, for example, received no coverage from *Time* during the period under study.[11] *Time* adopted the perspective of business, wrote about technological issues important to business, and helped to stabilize the meaning of computers around definitions acceptable to business.

As the meaning of the new technologies stabilized in *Time* in the mid-1980s, the language which the magazine used to describe computers changed. The language of magic and religion, which had first signaled that something new and wondrous was happening, and then played a role in defining the new technologies, began to decline. The metaphors of magic did not disappear from the pages of *Time*, but they did become less pronounced and more an invocation of already established meaning. As business successfully appropriated computers, the technological frame began to close.

---

[11] By my own count, there were about 160 books and articles on women and computing published in English during the period under study. *Time* made no mention of any of them.

Chapter Five

Faust's Bargain

FAUST: I've studied now Philosophy
and Jurisprudence, Medicine,
and e'en, alas! Theology,
from end to end, with labour keen.
Yet here, poor fool! with all my lore
I stand no wiser than before;
I'm Magister—yea, Doctor,—hight,
and straight or crosswise, wrong or right,
nigh ten years now, with many woes,
I've led my pupils by the nose...
and see that nothing can be known!
This knowledge cuts me to the bone.
Johann Wolfgang von Goethe
Faust, Part One

Wisdom brings up her own sons, and cares for those who seek her. Whoever loves her loves life, those who wait on her early will be filled with happiness. Whoever holds her close will inherit honour, and wherever he walks the Lord will bless him. Those who serve her minister to the Holy One, and the Lord loves those who love her.
Jesus Ben Sira
Ecclesiasticus 4: 11-15

Ask someone on the street what a myth is and she or he is likely to say "a story which isn't true." In advanced industrial society we like to think that we are scientific and secular, that in our enlightenment we have put myth behind us. Today we have banished myth to the specialized realms of literature and psychoanalysis and enthroned science and technology in its place as the pre-eminent forms of knowledge. We pride ourselves on our "reason," "common sense" and "realism." Yet at the same time we put the chiliastic visions of soothsayers on the bestseller lists and the media uses the language of magic and religion to speak of our most advanced technologies. We have become blind to our own metaphors, confusing our figures of speech with the way things are. We have become a literal-minded people with single vision, living in a demarcated world which turns out to be as illusory as the myths which we have repressed (cf. Bellah, 1970: 244). Technological mysticism is the return of the repressed. By not acknowledging our myths, we fall blindly into their thrall. We can never live without myths, but every culture has many stories and we have to take responsibility for which stories we tell. At the roots of technological mysticism lie some of the foundational myths of our society. One of its most central myths is Faust.

This chapter begins by looking at the nature of myth and its relation to science and technology in the modern world. I then argue that the Faust myth exemplifies the practice of science and technology which has become dominant in advanced industrial society. The paradoxes of the Faust myth are revealed in three "cautionary tales." The chapter concludes by contrasting Faust with another form of knowing, the Biblical Wisdom tradition as exemplified by Jesus Ben Sira.

Myth and Enlightenment

Myths are symbolic narratives. These are the stories which tell us who we are, what the world means, and what is required of us. In pre-modern societies myth functioned as explanation (Ricouer, 1967), they were stories of the origin of things and the doings of the gods. Some simply explained nature, as when thunder was construed as the sound of Thor's hammer or lightning as the wrath of Zeus. Even these etiologies of natural phenomena, however, were characterized by the breaking through of the sacred into the world (Eliade, 1963). Far more profound, of course, were those stories which related the origins of the world itself, of humanity, of birth and suffering and death. Mircea Eliade explains:

> Myth, in itself, is not a guarantee of "goodness" or morality. Its function is to reveal models and, in so doing, to give a meaning to the World and to human life. This is why its role in the constitution of man is immense. It is through myth...that the ideas of reality, value, transcendence slowly dawn. Through myth, the World can be apprehended as a perfectly articulated, intelligible, and significant Cosmos. In telling how things were made, myth reveals by whom and why they were made and under what circumstances. All these "revelations" involve man more or less directly, for they make up a Sacred History. (1963: 144-45)

In the great narratives of a society, its sacred history or what Frederick Ferré (1993) calls its mythic matrix, are found its fundamental values and its answers (or theodicies) to the unanswerable questions of life. Myths are not just "stories about the gods" but all those narratives which ground a people's identity, legitimize their social order, and give meaning to their lives. They are fundamentally, even if sometimes implicitly, religious.

Myths are *symbolic* narratives. Symbols are the basic building blocks of culture. Emile Durkheim recognized this more than eighty years ago. "A society can neither create nor recreate itself without at the same time creating an ideal," he wrote (Durkheim, 1915: 470). This ideal was not something extra or added on to "real" society, he claimed, because "society is not made up merely of the mass of individuals who compose it, the ground which they occupy, the things which they use and the movements which they perform, but above all the idea which it forms of itself" (Durkheim, 1915: 470).

The central question here is what Durkheim meant by "the idea which society forms of itself." As Robert Bellah has noted (1973), throughout his career Durkheim talked about *collective representations* (what today we call symbols) as "social facts," but late in his life he made an important distinction. In an essay on "Value Judgments and Judgments of Reality" in 1911 he wrote:

> If all judgments involve ideals we have different species of ideals. The function of some is to express the reality to which they adhere. These are properly called concepts. The function of others is, on the contrary, to transfigure the realities to which they relate, and these are the ideals of value. In the first instance the ideal is a symbol of the thing and makes it an object of understanding. In the second the thing itself symbolizes the ideal and acts as the medium through which the ideal becomes capable of being understood. Naturally the judgments vary according to the ideals involved. Judgments of the first order are limited to the faithful analysis and representation of reality, while those of the second order express that novel aspect of the object with which it is endowed by the ideal. This aspect is itself real, but not real in the same way that the inherent properties of the object are real. (Durkheim, 1924/1953: 95-96)

For Durkheim, the ideal takes two forms. Concepts represent an empirical reality and judgments about them concern how accurately that empirical reality is represented. But when he says "the thing itself symbolizes the ideal and acts as the medium through which the ideal becomes capable of being understood" he is talking about another form of reality altogether (Bellah, 1973: *li*), a reality constituted in and through the symbol and in which the symbols "transfigure the realities to which they relate." There is nothing one can "point to" when one speaks of God or justification or forgiveness. Instead they are, as Bellah called them, the "summary symbols that refer to the totality of being" which "individuals and societies have used to make sense of themselves and give direction to their action"

(1970: 202). They mediate a reality that lies in transcendent spheres of meaning. These are the central symbols of society.

The basis for this rests in the structure of symbols themselves. Alfred Schutz (1973: 301) argues that all symbols consist of a triadic relationship. First, there is the signifier, the sign or symbol itself. This can be anything—a word, an object, an image, music, etc.—accessible to our senses. Second, there is the signified, that to which the signifier points, its meaning. And finally there is an interpreter. All three elements are necessary if a symbol is to convey meaning. Obviously, the signifier is necessary or there is no symbol. Absence of the signified would be like hearing a word in a language we do not understand, or seeing a curious object the purpose of which is unknown, we perceive that something is there (the signifier) but we do not know its meaning. But the interpreter is just as necessary or the symbol stands mute. There is no such thing as meaning independent of people (Lakoff and Johnson, 1980: 184); they are an intrinsic part of the symbolic relationship.

Schutz's insight has several implications. The dichotomy between subject and object is broken down. Symbols are objectively real, they exist prior to us and apart from us and we cannot make them mean whatever we please, but at the same time our subjectivity as interpreters is part of every symbolic relationship. So symbols are neither subjective nor objective, but relational (Bellah, 1970: 202). Interpretation of symbols is therefore neither an attempt to substitute an objective "true knowledge" for a subjective "false consciousness," nor to reduce experience to a theoretical formula, but instead is an attempt to communicate and understand. In Jürgen Habermas's words, when we engage in interpretation we attribute "epistemic authority to the community of those who cooperate and speak with one another" (1991: 19). What constitutes our reality, then, is our community of discourse.

Because they are inherently relationships, symbols are also able to mediate what Schutz (1973) calls different *spheres of meaning*. One of the properties of symbols is that they have more than one meaning (Ricoeur, 1970). A single signifier may point to a number of signifieds, as when a word has more than one meaning. The English word "fire," for example, may mean a rapid process of oxidation producing heat and light, or it could be a command to soldiers to discharge their weapons, or it could mean that someone has lost her or his job. A specific meaning can be determined, but only in the context of larger discourse (in the case of a word, a sentence or paragraph). When a person interprets a symbol these multiple meanings are put in conjunction and may "bleed" or "merge" into each other, creating the possibility for new spheres of meaning to emerge. Various types of symbols (metaphors, similes, metonymies, etc.) structure these conjunctions in different ways, opening different possibilities. As a consequence, the number of meanings for any symbol is open. The full range of possible meanings may exceed the knowledge or awareness of any given interpreter. Even if it were possible for any interpreter to find out all a symbol means and has meant, the possibility of new meanings appearing could never be closed off.

Myths are symbolic *narratives.* They are stories which are told, which are related through our language. We can never separate meaning from our discourse. As the philosopher Richard Bernstein said: "Language is the medium of all understanding and all tradition. And language is not to be understood as an instrument or tool that we use; rather it is the medium in which we live"(1983: 145). We cannot separate experience as such from our linguistically mediated understanding of it (Bellah, 1970: 202). Our language and the symbols which make it up are grounded in experience. Linguists George Lakoff and Mark Johnson demonstrate this in their analysis of metaphor. They argue that "no metaphor can ever be comprehended or even adequately represented independently of its experiential basis" (1980: 19 italics omitted). Orientational metaphors, for instance (happy is up, sad is down; more is up, less is down; good is up, bad is down; etc.), are grounded in our physical spacialization (pp. 14-21). As mammals which walk upright, they claim, we experience "happy" or "good" as "up," their opposites as "down." But at the same time as metaphors are grounded in our physical experience, they structure that experience into a consistent conceptual system. Our experiences, indeed human thought processes themselves, are fundamentally metaphorical (p. 6). Lakoff and Johnson explain:

> What we call "direct physical experience" is never merely a matter of having a body of a certain sort; rather *every* experience takes place within a vast background of cultural presuppositions. It can be misleading, therefore, to speak of direct physical experience as though there were some core of immediate experience which we then "interpret" in terms of our conceptual system. Cultural assumptions, values, and attitudes are not a conceptual overlay which we may or may not place upon experience as we choose. It would be more correct to say that all experience is cultural through and through, that we experience our "world" in such a way that our culture is already present in the very experience itself. (1980: 57)

There is no separation between our experience and our language. We do not have experience apart from our symbolizations of it any more than we can step outside of society in order to interpret it. As Robert Bellah says, "we can only think in symbols" (1970: 202).

The modern world rejected myth for science and technology but in doing so it turned science and technology into myth. At the beginning of the modern era, Francis Bacon set the agenda for the new science as the translation of knowledge about nature into power over nature through the use of technology. In order to have power over nature, people had to overcome what Bacon called "idols"[1] in order to see reality. "Truth is to be sought for," Bacon wrote in the *Novum Organum*, "not in the felicity of any age, which is an unstable thing, but in the light of nature and experience, which is eternal" (1620/1960: 55). Observation of nature was to replace speculation, empirical facts were to replace symbol and myth. "To resolve nature into abstraction is less our purpose," he said, "than to dissect her into parts" (1620/1960: 53).[2] Only an empirical science would be strong enough to break myth and force nature to give up her secrets.

Bacon set the agenda for generations of scientists. The Enlightenment boasted that they had abandoned tradition and myth for reality. Science was the royal road to truth. A *scientific worldview* soon emerged which went beyond the scientific method as a tool of analysis to challenge religion as an overarching system of meaning (Ferré, 1993). What began in Bacon's day as a conflict between scientific discoveries and the authority of the church,[3] grew by the eighteenth and nineteenth centuries to an attack on the foundations of religion itself (as in the astronomer Laplace's famous dictum, "I have no need for that hypothesis"). The irony of the Enlightenment, however, is that in destroying traditional myths, science was elevated into their place. Any worldview is, by definition, a symbolic narrative, a mythic matrix which explains "reality." As science became a worldview it began to take on the roles and functions of religion. Enlightenment became myth (cf. Horkheimer and Adorno, 1944/1972).

There is no access to "reality" which is not mediated by language. For all their pride in finding "facts," scientists and philosophers had overlooked the fact that their own language is metaphorical (Gerhart and Russell, 1984). As Bruce Gregory explains: "The observations with which physicists compare their predictions are not some mute expression of the world. They are symbolic and gain their meaning and value in a system of interpretation. No experiment...has any meaning at all until it is interpreted by theory" (1990: 191). Empirical observation can never be separated from the language in which it is imbedded. For example, Evelyn Fox

---

[1] Bacon saw four types of idols as barriers to the new science. Idols of the tribe were beliefs commonly held by the people, idols of the cave were based on idiosyncratic interpretations by individual scholars, idols of the marketplace were those imposed by words and names and idols of the theatre were imposed by philosophical systems. His critique was specifically directed at the religion, theology (especially scholasticism) and philosophy of his day. For an interesting commentary, see Leiss, 1990.

[2] Note the image here of an anthropomorphized female nature in the torture chamber. Evelyn Fox Keller (1985) analyzes the misogynistic language of Bacon and other early scientists.

[3] Remember that Galileo was put on trial not for the *content* of the Copernican model but because he continued to teach it when ordered to stop.

Keller (1992) shows how the language of molecular biology, which speaks of "selfish" genes engaged in evolutionary "competition," is both androcentric and grounded in liberal capitalist political economy. For all their reductionism, cultural perceptions of reality creep into their science at the pre-theoretical level. So scientists, too, can only think in symbols. The world described by science is a symbolic reality.

Therefore science is not an escape from symbol and myth, but in pretending that it was the Enlightenment way of thinking opened the door for myth to return *sub rosa*. Technological mysticism is the return of the repressed.

We cannot live without myths, but by not acknowledging our myths, we fall blindly into their thrall. When we forget that we do not have experience apart from our symbolizations of it, we confuse our figures of speech with reality. No one is more deluded than the person who thinks he or she is "just being realistic." We become people of single vision who live in a literal and circumscribed world which turns out to be as illusory as the myths which we have repressed (Bellah, 1970: 244). For a complex and multidimensional world mediated by symbols, we too often substitute our own blinkered vision, then loudly trumpet our "success" because our theory "works." Of course, such "success" is usually measured only in the short term and by "objective" indicators and standards which are themselves socially constructed as part of the relations of ruling (D. Smith, 1990) and which all too often hide the real social and environmental costs. But even more serious is the loss of meaning. Cost-benefit ratios and leading economic indicators cannot tell us who we are, or why are we here, or what is right for us to do. But people need meaning as much as they need food and shelter. So when we deny our own myths, we end up being ruled by those of others (Midgley, 1992: 13). There is always a huckster or zealot willing to peddle transcendence to those starved for spirituality or meaning. Ideologues today are particularly fond of Margaret Thatcher's dictum "there is no alternative." From the drumbeat of identity politics (all the infinite variations on the theme "we are the chosen people"), to the voodoo of economic theory, to the hidden subtexts of reductionist science and fundamentalist religion, they pronounce their nostrums to be "the way things are." People who have forgotten their own stories and traditions are an easy mark.

So we can never live without myths, whether we acknowledge them or not. But just because we need myths does not mean that all myths are equally good. We also have to remember that myths are inherently ambiguous and we can never escape their ambiguity. Myths ground identity, but identity can at times be oppressive or stultifying. They legitimize the social order, but that order may be tyrannical or destructive. The twentieth century has seen more that its share of tragic and brutal attempts to create meaning.

Fortunately, every culture has many stories. Some myths may be destructive or oppressive, but in every society there are alternatives. Many myths are fostered by the powerful; others are found on the margins of society. Even within one myth there are many ways the story can be told. Because an interpreter is an essential component of every symbol and because the number of meanings for any symbol is open, each telling of a myth is a narrative reweaving. Every time a myth is recounted its symbolic building blocks are reassembled by the interpreter in a new pattern. A storyteller is like the conductor of an orchestra who, while following the score, may accentuate the strings or hush the brass.

So there are *always* alternatives. When some said knowledge is power, others replied with tales of Victor Frankenstein or Captain Nemo or Dr. Strangelove. We cannot avoid telling stories, but we have to take responsibility for which stories we tell, and how we tell them.

The Faust Myth

Faust is the story of the man who sold his soul to the Devil in exchange for knowledge and power. Apparently there actually was a Doctor Faustus. As best as can be reconstructed,[4] Johann (or Georg) Faust was born about 1480 in central Germany and died in 1540. In all accounts he was an educated man, perhaps known to Luther and Melanchthon, but he devoted his life to wandering about casting horoscopes, performing magic tricks, and boasting of his exploits. Many stories grew up around Faust and many traditional folk tales were worked into the legends (Haile, 1965). By the mid-sixteenth century several collections of these stories began to appear as chapbooks in the Protestant parts of Germany. In 1587 Johann Spiess published the Faustbook, which (through a bad translation), became the basis for Christopher Marlowe's *The Tragicall History of the Life and Death of Doctor Faustus.* Since then scores of variations have appeared, from crude puppet plays to such monuments of literature as those of Goethe and Thomas Mann.

The importance of Faust to Western culture cannot be overestimated. It is a persistent theme in modern art and literature. Jeffrey Russell reports that: "The figure of Faust is—after Christ, Mary, and the Devil—the single most popular character in the history of Western Christian culture...the story is a lietmotif of Western art for half a millennium" (Russell, 1986: 58). H. G. Haile adds that: "We can call Goethe's drama *the* representative work of our modern civilization,

---

[4] Scholars, as usual, disagree over the evidence. See Berghahn, 1987; Brough, 1994; Haile, 1965.

because it still seems to be our final statement of faith in man's inborn drive: 'The man who keeps on striving, he can be saved'" (Haile, 1965: 14). Faust is a foundational myth of advanced industrial society, an exemplar of our practice of science and technology. Faust's striving echoes through modern virtuosity and economic values. With such a complex story and such a large literature, I cannot explore all its ramifications here but I would like to look at one theme, as expressed in the Faustbook,[5] Marlowe and Goethe, which I think has particular bearing on our understanding of technology today.

In all three of these versions, Faust was a scholar who, having reached the limits of human learning, turns to magic in an insatiable quest for knowledge. Goethe most clearly expresses his unquenchable striving:

> I've studied now Philosophy
> and Jurisprudence, Medicine,
> and e'en, alas! Theology,
> from end to end, with labour keen.
> Yet here, poor fool! with all my lore
> I stand no wiser than before;
> I'm Magister—yea, Doctor,—hight,
> and straight or crosswise, wrong or right,
> nigh ten years now, with many woes,
> I've led my pupils by the nose...
> and see that nothing can be known!
> *This* knowledge cuts me to the bone.
>
> ([part one, lines 354-366] 1962: 59)

Turning to the spirit world for the answers he cannot find in the material one, Faust makes a pact with the Devil in which Mephistopheles is (in the words of the Faustbook) "charged with informing and instructing me, and...to be subservient and obedient to me in all things" (Haile, 1965: 33). Faust's ambition was to translate knowledge into power, to, in Marlowe's version:

> ...be great emperor of the world,
> And make a bridge thorough the moving air,
> To pass the Ocean with a band of men.
> I'll join the hills that bind the Afric shore,
> And make that country continent to Spain,
> And both contributory to my crown.
>
> ([I, iii] Marlowe, 1966: 12)

---

[5] I use H. G. Haile's (1965) translation of the Wolfenbüttel manuscript [1580], the oldest existing version. Spiess used this manuscript for his published version.

Faust wanted no limits. Magic was to be the technical means of achieving his vision. Goethe, having turned Faust into a Romantic hero, substitutes for the pact a wager that Faust's soul would be forfeit should he ever be satisfied, a variation which permits Faust to be saved at the end. The two earlier stories conclude with Faust's grisly death and damnation at the expiry of his contract.

The Devil does not live up to his part of the bargain. Faust gets magical knowledge and power, but it does not measure up to his expectations. "Much had been promised by his spirit, but much had been lies, for the Devil is the spirit of lies" says the Faustbook (Haile, 1965: 86). Faust begins with great ambitions, but very quickly Mephistopheles turns him aside into a life of debauchery (Levin, 1966). For the price of his soul, the Faustbook and Marlowe have Faust performing tricks, petty exercises in power that only satisfy his lusts. In Goethe, Mephistopheles delivers more but twists every exercise of power in a way Faust had not intended. In none of the stories is Faust satisfied with the results of his bargain. The tragedy of Faust is that of "the scientific libertine who gained control over nature while losing control of himself" (Levin, 1966: 145).

This theme is a potent metaphor for our technology practice, in three ways. Like Faust we have renounced limits to our knowledge, and today we use our technology to probe from the limits of atomic structure and the genetic code to beyond the limits of the galaxy itself. Like Faust we have sought control to the point of hubris, an overweening pride in our own power. And like Faust, we have begun with noble intentions which have gone unfulfilled.

Like Faust we have renounced limits to our knowledge. Prior to the coming of modern society limitations on what could and should be known were widely accepted. Even Plato advocated placing restrictions on inquiry.[6] The question of whether or not knowledge is *good* is a non-question today. Occasionally a few people will raise questions about how knowledge will be *used*—should we have built the atomic bomb; what are the consequences of genetic engineering—and there are frequent disputes over the allocation of research funds, but virtually no one questions the search for knowledge itself. Pushing the limits of knowledge is seen as intrinsically good. Restricting inquiry is believed to be an unacceptable violation of our freedom when done by others and an act of moral and intellectual cowardice if we impose limits upon ourselves.

Faust has not been alone in equating knowledge with power. Just as Faust abandoned the holistic scholasticism of his past when he turned to magic, modern virtuosity and economic values abjure limits in their search for ever greater mastery. The ambition to dominate and control are not

[6] See, for example, the infamous "nocturnal council" in *The Laws*.

unique to today's world, but what characterizes moderns is what Frederick Ferré calls the *control syndrome*, "the supposition that we have or should aspire to absolute autonomy unbalanced by holistic mutuality, whether in the short term or the long" (1993: 169). The ancient Greeks, for instance, were as acquisitive as any and more inquisitive than most, but they recognized limits. Their chief sin was hubris, a prideful overreaching which was sure to bring nemesis in the form of retribution from the gods. Today, like Faust, we acknowledge no limits to our power and control. A common attitude is that if something *can* be done it *should* be done (often with the insinuation that if we don't, someone else will). For Ferré, "the control syndrome is what has brought humankind to its present parlous condition" (1993: 170). Matching our desire to master nature with our need to master our neighbour, modern society has been built upon economic and political dominance hierarchies which ignore reciprocity and holism, see competition and greed as virtues, countenance limitless exploitation, and are continuously at war or preparing for war.[7] Our tragedy is, like Faust, to have gained control over nature and others while losing self-control.

And like Faust, we have begun with noble intentions which have gone unfulfilled. We have seen the diabolical face our technology can take—at Bophal, at Chernobyl, at Hiroshima—but perhaps even more enervating is the daily anomie of life in a society in which there are "fifty channels and nothing to watch." We have enormous power but very little sense of direction. We are masters without meaning. We have become, in Max Weber's famous words, "Specialists without spirit, sensualists without heart; this nullity imagines that it has attained a level of civilization never before achieved" (1905/1958: 182). Some might even go further and see in the growing environmental crisis an echo of the duration of Faust's contract. Faust was given power for twenty-four years; then his soul was forfeit. Some deep ecologists might say that industrial society has run its course and, like Faust, our time is nearly up. While this may be too pessimistic, Faust's bargain still lies at the heart of computer and related technologies, as can be seen in the following cautionary tales.

Some Cautionary Tales

Just as Mephistopheles gave Faust knowledge and power, but always twisted it in a way that Faust did not anticipate, the culture of technology promises more power and control than it delivers. Certainly today we in the West have more of both than any previous period in history. But as with Faust, our power does not quite work out as we think it will. Just

---

7 This is only a capsule summary of Ferré's criticism of modernity, which will be explored more fully, along with that of several others, in chapter six.

consider a few stories about technology in the news in the mid-summer of 1996.

At the Atlanta Olympics, a computerized data system which was to be a showpiece for IBM turned out to be full of bugs and did not perform as expected. In Oregon, someone took personal information from the Department of Motor Vehicles about every driver in the state and posted it on the Internet. On two consecutive days, electrical power failures blacked out portions of six western states and Alberta. In Canada, a serial rapist and murderer was caught after ten years when police finally entered data on his crimes into a national police data bank. It seems that data entry involves filling out a thirty-six page form, and the understaffed police in Ontario had not gotten around to doing it any sooner. The system crashed at America On-Line, knocking one of the largest service providers off the Internet. In Alberta, a cut of twenty grain cars somehow rolled out of a CN yard and down the mainline, evaded prompt detection by the computerized traffic control system and crashed headlong into a freight train, killing three people. And an essay in *Time* magazine commented that the Internet has so far failed to deliver on promises of democracy (Wise, 1996).

This list is not extraordinary, or even complete. The failure of technology to perform as we expect it to is both a normal and routine event. Like Faust's pact with the Devil, our technology fails to deliver the promised limitless power and wealth. Three recent books explore the paradoxes of Faust's bargain. They are cautionary tales.

In *Why Things Bite Back: Technology and the Revenge of Unintended Consequences,* Edward Tenner (1996) describes what he calls revenge effects, the "ironic unintended consequences" (p. 6) of our technology. Revenge effects are neither side effects nor trade-offs. A side effect is an additional consequence of the use of a technology, as when fat substitutes cause diarrhea, while trade-offs are the price we pay in order to get certain benefits, as when mandated auto safety features such as air bags increase the cost of a car. Revenge effects are, as Tenner puts it, "the tendency of the world around us to get even" (p. 5). They are ironic in that they exacerbate the problem the technology was intended to solve. For example, revenge effects occur if automobile air bags injure small children. Revenge effects are systemic. They emerge out of the interplay of technical, organizational and cultural aspects of technology. They also may not be visible if only the narrowly technical is examined, which is why policies based on a technical fix so often end up appearing blind in retrospect.

Tenner sees five kinds of revenge effects. The first are *rearranging effects*, which occur when technology does not solve a problem but simply moves it around. For example, business in the 1980s and 1990s spent billions of dollars to computerize their offices in an attempt to increase productivity,[8] yet overall increases in office productivity have been negligible (estimates range from 0.02 percent to 2.2 percent [p. 188; see also: Gibbs, 1997; Landauer, 1995]). Why? Hidden costs are a major reason. Computers often replace low-wage workers but substitute for them fewer but very highly paid workers who are required to program, service and maintain the machines (p. 191). After computerization, high-priced executives spend much more of their time sending their own correspondence (and e-mail). The most technically proficient employees (who would be the most productive with the new technology) spend enormous amounts of their time doing informal peer support and training (pp. 197-98). Countless hours can be spent unproductively "surfing the net" or playing computer games. So while the information services budget may record a cost of $2,000 to $6,000 for each workstation, actual costs may be three times that (p. 198).[9] Computers have not reduced costs; they simply moved them around.

The second form of revenge effects are *repeating effects*. These occur when "labour saving" technology results in using up all the time saved by doing the same thing more often. It has long been noted that the amount of time spent on housework has not significantly declined, in spite of the introduction of household appliances. Every technical advance has been matched by an increase in the cultural standard of cleanliness. The same is true in the office. Word processors give anyone the ability to produce typographically perfect documents. Consequently, perfect documents have become the standard, and typos which would have been overlooked two decades ago now draw rebuke. Documents are edited and reedited, since easy revision means that revision is required. The end result is that just as much time is spent preparing documents as before. And to gain further revenge, computers have created a wide range of repetitive stress injuries to slow us down even more (pp. 161-83).

In 1983, my first computer was a top-of-the-line machine. It had 64K of RAM, a dual floppy disk drive and a monotone monitor. I expanded it to 256K RAM, but could not envision any software I would need which would require one of the new (and very expensive) Winchester hard disks. Things have changed. I recently downloaded a game *demo* which required forty-five megabytes of hard disk space and eight megabytes of RAM to open. Of course, in part this shows the incredible advances in computer technology in the past fifteen years, but it also demonstrates what Tenner means by *recomplicating effects*. These make a problem

[8] Gibbs (1997: 82) reports that the 1996 information technology budget was $500 billion in the U. S. and $1 trillion worldwide.

[9] Others claim the actual costs are even higher. Halfhill (1997: 70) estimates the five-year cost of a workstation equiped with Windows 3.1 at $44,250, with Windows 95 at $38,900, and Windows NT at $38,400.

more complicated instead of simpler. In the 1950s, Detroit kept producing automobiles which grew larger and flashier, with more chrome and bigger tail fins (and bigger price tags) and which were planned to be obsolete in two years. Then European and Japanese compact cars arrived in the 1960s and 1970s and the Big Three auto companies were thrown into crisis. In many ways, Silicon Valley today is like Detroit in the 1950s. Memories are larger and processors are faster but software has grown correspondingly to match it. Code inflation has meant more memory is tied up in overhead. For example, Microsoft Windows 3.1 with MS-DOS required 20 mega-bytes of hard disk space; today Windows 95 uses up 60 megabytes (p. 194). More "bells and whistles" eat up memory and complicate programs. Microsoft Word, for instance, has grown from 311 commands in its 1992 version to 1,033 commands in Word 97 (Gibbs, 1997: 88). Meanwhile the plethora of operating systems, changing machine architecture and multiple, incompatible "standards" means that retaining and storing data becomes a complicated problem. Archives, for instance, are rapidly becoming museums of information technology in an attempt to maintain records stored in obsolete formats. Tenner comments:

> Programmers and developers are understandably working for the future, preparing for the next generation of machines. Most users are living in the past because either their budgets or their bosses won't let them upgrade hardware to the current standard. Meanwhile the planned obsolescence of software means that earlier versions of major programs may no longer be supported. (p. 194)

Users are chronically short of computing power and forced to upgrade. And unlike used cars, which still have resale value, older computers are virtually worthless (p. 185-86).

The fourth form of revenge effects, *regenerating effects*, make the original problem worse. One of the most touted promises of computerization was the "paperless office." Costs would be reduced, it was prophesied, because when everyone was on a computer network there would no longer be a need to put anything on paper. Since computerization, paper use has gone up 250 percent. Not only do many printers waste paper, but everyone wants their own hard copy of whatever was sent, revision after revision of documents are printed out, and the original document has achieved a near sacred status—to be carefully protected and filed away while everyone works from their own copy. Because producing more copies of a document is easy, more copies are produced and paper use escalates.

Recongesting effects are the final form of revenge effects. Whatever new "space" is created by technology quickly becomes fully occupied. The best example of this is the "info glut"—the massive and accelerating increase of information available to the user. Finding valuable information amidst all the "noise" can be a trying experience. A story posted to the newsgroup *comp.risks* recounted the sad tale of someone who tried to search for the home page of "Kirk Enterprises" on the World Wide Web, only to be inundated by references to "Star Trek". The information explosion has grown to such an extent that few among even the narrowest of specialists can hope to stay fully informed. For example, a few years ago, I presented a paper on "computer literacy" at a seminar in India (Stahl, 1994). Based on a computer search, I calculated that if all the articles on computer literacy published in English were immediately available and I read on average one article every two hours, eight hours a day, five days a week, fifty weeks a year it would take me 1.39 years just to get caught up. This assumes I did nothing else than read journal articles and that no more would be published while I was working on the backlog! The Internet promises to put the world's information at our fingertips, but recongesting effects mean that we are likely to be buried in an avalanche of data.

Revenge effects have had several consequences, according to Tenner (other than negating the more utopian speculations on what technology will do). Few of our problems disappear, but many shift from being acute to chronic. For example, since the nineteenth century there has been a significant decline in worker on-the-job deaths. This has been matched by an increase in chronic injuries, especially back and repetitive stress injuries (pp. 161-83). The greater power and complexity of today's technology requires increased vigilance, which raises questions of freedom and privacy. Increased vigilance has also meant that the frequency of disasters has declined but, ironically, when they do happen they tend to be worse, just as routine fire suppression makes forest fires worse when they finally do happen. For instance, in the nineteenth century steam boilers exploded regularly, often killing a few—or even a few dozen—people each time. Today's nuclear reactors are far safer, but when one of them melts down thousands die. As Tenner concludes "the safer life imposes an ever-increasing burden of attention" (p. 254).

A second cautionary tale is Lauren Ruth Wiener's *Digital Woes: Why We Should Not Depend on Software* (1993). In 1968 the term *software crisis* was coined to describe the difficulty systems engineers were having in developing reliable systems (p. 31). Nearly thirty years later, Wiener reports, the crisis is still very much with us. Indeed, she argues that "it's time we recognized that this is not a crisis, it's a situation: software has bugs. It is in its nature to have bugs, and that fact is unlikely to change soon" (p. 193). Software is inherently unreliable.

Software unreliability stems from a number of interrelating factors. Computers manifest a duality akin to body and soul, she says (p. 38). They exist both as physical systems (the hardware) and as logical systems (the software). Problems can arise out of either and out of the interaction between the two. Bugs can occur in the technical aspect of writing code. Unlike other engineers, for whom physical laws are a constant, software engineers can take nothing for granted—they must specify *everything* (p. 45). They must anticipate every contingency in what may be a large and complex system, how their system will interact with other systems, and all the ways people may use and abuse it. Since computers have no "common sense," the most ordinary tasks must be spelled out in complete detail, which itself may be a formidable undertaking (when the U.S. Occupational Health and Safety Administration tried to describe how to climb a ladder the text ran to several pages). All of this must be typographically perfect (the *Mariner 1* space probe failed when one bar (¯) was omitted in a program running to thousands of lines of code [pp. 4-5]). And every attempt to fix a bug may generate a new one since bugs arise from both errors in the code and from the interaction of different programs and systems with each other.

The organizational aspects of technology practice also generate bugs. Market pressures mean that software development is too often given insufficient resources and too little time. The rapid pace of technological change means that little experience is gained with one system before the next one comes along. The structures of most corporations are poorly suited to software development. Wiener explains: "Dominance hierarchies may not be the *absolute worst* mechanism for getting some sort of work done, but they aren't particularly suited to software development. Invention comes unbidden; subtle stubborn fact cannot be commanded; technical details must not be ignored" (p. 78).

Corporate structures, usually run by managers with little technical expertise, are often at odds with the subculture of programmers, who frequently reject hierarchy and the authority of those lacking technical prowess (p. 79). The result is lack of communication, frustration, and high rates of project failures and programmer burnout. Complicating all this is the absence of an adequate model for software development (pp. 82-107). The various steps of software development—specifying requirements, design, programming, and testing—are interactive rather than sequential, that is, one cannot complete one step and move on to the next. Each involves reference to, and revision of, the others.

Finally, our culture is so dazzled by the possibilities of digital technology that its limits are often overlooked. Wiener explains:

> We hear a lot about the benefits of proposed systems, and many of them are beneficial, certainly. But we seldom hear what we are risking or get an accurate picture of the cost. Public discussion tends to be sparse and one-sided, often dominated by those thrilled to breathlessness by the thought of

> building these systems or collecting the profits from them. They tend to describe benefits in glowing, unrealistic terms. (p. 144)

In one of her many examples, she warns of overreliance on computer models and simulations as decision-making tools (pp. 160-66). Any model is, by definition, a selection and interpretation of reality, not reality in miniature. The reliability of its output is a function of the quality of its inputs, or as the old saying goes, "garbage in, garbage out." Of course, it is rarely easy to determine the quality of inputs, since these are selected by the skill, judgment, assumptions, biases, and ideology of the programmers. As long as these limitations are remembered, models can be useful. The danger arises when the decision makers do not understand the model's inherent limitations. For example, the debates over global warming have been bogged down in part because of conflicting models of climate change (pp. 163-64). Can policy makers take the risk of waiting for perfect information (which will not be forthcoming, at least until it is too late)? Or do some pick the model which confirms their biases as an excuse for their action (or inaction)?

Thus, in every aspect of technology practice, software is unreliable. Fortunately, for most software this may not matter very much. However in *mission critical* and *safety critical* software, on which large amounts of resources or human lives depend, extraordinary measures must be taken to ensure reliability (pp. 112-41). This can be done, at great cost and up to a certain degree, but no system will ever be completely fail-safe. She concludes by asking; "when we build a system, we should ask ourselves whether we will be worse off when it fails than we would have been if we hadn't built it" (p. 178). The question of whether or not a technology is appropriate applies as much to computers as anything else.

Bruno Latour's *Aramis or the Love of Technology* (1996) is an unusual book, part novel and part documentary. Through the device of two narrators, a professor and his research assistant who are often in conflict, Latour tells the story of Aramis, a personalized rapid transit system for Paris begun in 1969 and canceled in 1987 after the expenditure of half a billion francs.

Aramis began with a grandiose vision in the 1960s for a system which would combine the convenience and privacy of the automobile with the speed and efficiency of mass transit. What emerged from the drawing boards was a system of hundreds of small cars shuttling people from scores of stations into the core of Paris. As envisioned, a person would walk to one of the many stations located only a few hundred metres apart and summon a car. Once aboard, that car would join with others to form a train moving at high speed and deposit the passenger at any other station. The entire system would be operated by computer.

The technical challenges of Aramis were formidable. In combining the virtues of automobiles and mass transit, it also combined the disadvantages of each. Physically moving large numbers of small cars rapidly and safely to where they were needed as traffic demands shifted throughout the day was well beyond the ability of any control software yet written. In order to move large amounts of traffic swifty the cars must be formed into trains but, unlike railroads and subways, the cars could not be mechanically coupled together if speed and flexibility were to be maintained. Aramis was unable to surmount these challenges.

While telling the tale of Aramis's failure, Latour also demonstrates his own version of constructivism, a variety of semiotics which he calls *actor-network theory*. The result is a complex interweaving of actors (including Aramis itself), none of which speak as the "subject" of the book. It is from this dialogue that the reasons for the failure of the project emerge. No one loved Aramis. The initial vision was enough to mobilize politicians, bureaucrats and entrepreneurs to begin the project but no one emerged as an *heterogeneous engineer* (in Latour's jargon)[10] who could bind together a network which would hold. Aramis was caught between the Paris METRO (who had a subway to run), the prime contractor (who had a competing system in Lille), and the bureaucrats of the Ministry of Transportation. There was no one like Admiral Rickover, pounding on desks to overcome navy inertia in order to get his atomic submarines. When politicians asserted themselves, the project moved, when they lost interest (or were replaced) the project languished. Each of the major institutional players was eager to get into production but no one was interested in funding research. Consequently, in spite of seventeen years and huge amounts of money, the basic technological challenges were never overcome. No more than three cars were ever held together by non-mechanical coupling and the control system could never guarantee safety. The problem was that all the players thought that technology was autonomous, and that once begun it would take care of itself. As Latour has Professor Norbert conclude:

> They really succeeded in separating technology from the social arena! They really believe in the total difference between the two. To cap it off, they themselves, the engineers and the technologists, believe what philosophers of technology say about technology! And in addition, research for them is impossible, unthinkable; its very movement of negotiation, of uncertainty,

---

[10] Others call such persons *product champions*. These are enthusiasts within an institution who challenge organizational inertia to promote an innovation. Some claim innovation is unlikely without a champion. See Chandler et al., 1992; Fisher, Hamilton, McLaughlin & Zmud, 1986; Schon, 1963; Westrum, 1991.

> scandalizes them. They throw money out the window, but they think research means throwing money out the window. (p. 287)

Because no one loved Aramis, its network came apart and the project died.

The Moral of the Story

Each of these three tales advises us to follow the path of wisdom rather than to be caught up in Faust's bargain. Compared to the utopian soothsayers discussed in chapter two or the magic language of the news media, Tenner, Wiener and Latour are sober, and sobering. Each tells an alternative story to the Faustian subtext of the dominant technology practice. Together they provide an antidote to technological mysticism. Each, in their own way, argues for a systematic approach to technology that combines the social and technical into a seamless web.[11] They offer a wise practicality that in theological language would be called Prudence.

The first step towards Prudence is to distance ourselves from the mythologies of technological mysticism. The paradox of Faust's bargain is that when we talk about technology as if it were autonomous we, like Faust, give up our own autonomy for the illusion of control over nature. When we blindly equate knowledge with power, reason turns into its opposite. We speak of technology as magic while we grasp for ever greater control, seeking limitless power with only partial understanding. Latour responds with a sarcastic curse: "People who talk about autonomy, irreversibility, and inertia in technology are criminals—never mind the purity of their motives. May the ashes of Chernobyl, the dust of the *Challenger*, the rust of the Lorraine steel mills fall on their heads and those of their children!" (1996: 86). If we are to be prudent, we must first debunk the hubris of control.

For Tenner, the power that technology gives with one hand it takes back with the other. Prudence requires a clearheaded assessment of the state of our technology, its strengths and its limits. It also requires patience and heedfulness. He writes: "It comes as no surprise that computing, like most other forms of contemporary technology, is neither a miracle weapon nor a dud, but a set of tools that need constant attention and maintenance" (1996: 199). Faust's bargain means that our empowerment brings with it servitude, that our freedom from fate has been purchased at the cost of constant vigilance and precaution.

Instead of technology's universal efficacy, Wiener cautions us to see its limitations. Instead of its autonomy, she emphasizes our role in

---

[11] Latour, of course, would substitute his own terms, claiming "social" and "technical" are too abstract. Actor-network theory does not assume "society" but tries to explain it. See 1996: 133.

making, and taking responsibility for, technological decisions. Most of all, we have to look at computers systematically as only one component in society and build those systems which are appropriate. She concludes:

> Digital technology is a useful tool, but the hammer doesn't build the house by itself. It must be wielded with purpose and judgment. Digital technology is a fascinating tool, but fascination is not an end in itself. We can fall prey to foolish fascinations. It takes a clever question to turn data into information, but it takes intelligence to use the result. Intelligence can create systems of enormous complexity, but it takes wisdom to determine which ones are worth the trouble. (1993: 209)

For all three of them, we must foresake Faust's bargain and follow the path of wisdom.

Faust or Ben Sira?

Are there alternatives to Faust's bargain, or have we, too, gained power only at the price of our souls? What other stories do we have to tell? Wiener calls for wisdom in the development and use of technology, but what does that involve and how does it differ from the Faustian search for unlimited knowledge and power? One place to look is the Biblical Wisdom tradition. The Wisdom tradition is not strictly parallel to Faust in that it is not a myth in the same way that Faust is, but it *is* a symbolic narrative which could become an alternative to Faust's bargain.

A good exemplar of the Wisdom tradition is Jesus Ben Sira, who lived and wrote in Jerusalem about 180 B.C.E. His book, called *Ecclesiasticus* in the Bible, is the only major work of Wisdom literature for whom we know much about the author. He wrote at a time when Israel was occupied by the Seleucid (Greek) empire and was a firm opponent of Hellenization.[12] Perhaps ironically, his book is primarily known from the Greek translation made by his grandson in Egypt in 132 B.C.E. Typical in style and format with the rest of the Wisdom literature, Ben Sira goes further than most identifying Wisdom (which he personified as female) with the Law and with the sacred history of Israel. He was the last author in the canonical Wisdom literature and is important to us as an exemplar and representative of that tradition. James Crenshaw analyzes Ben Sira and the other Wisdom books of the Hebrew Scriptures and provides the basis for an interesting contrast with the Faust myth.

Crenshaw defines Wisdom as "the reasoned search for specific ways to assure well-being and the implementation of those discoveries in

---

[12] The Seleucids tried to consolidate their heterogeneous empire by assimilating their subjects to the Greek language and culture. This policy would led to the Maccabaean revolt a few years after Ben Sira wrote his book.

daily existence" (1981: 24). Biblical Wisdom has several characteristics. First, Wisdom "consisted of a conviction that being wise meant a search for and maintenance of order" (p. 19). This was a theology of propriety and limits. Everything had a time and place, and the wise were those who could discern when any deed or word was proper and appropriate. Contrast this with Faust, for whom there were no limits. Second, Wisdom was characterized by moderation and restraint. Things which were good in themselves, even piety, when "carried to excess became evil in their consequences" (p. 20). The wise were prudent, in contrast to Faust whose life was devoted to excess. Third, Wisdom had its foundation in the fear of God. This was not the same as blindly following religious doctrine, for even the Torah was subject to commentary, but rather the belief that God's Wisdom was the beginning of all human knowledge. Faust, of course, sold his soul to the Devil. Fourth, the teachers of Wisdom "appealed to a sense of self-interest and relied upon a capacity to reason things out" (p. 21). The sages of ancient Israel employed a rich repertoire of literary styles and rhetorical devices because they believed it was necessary to *persuade* the people of the rightness and benefits of their teachings, rather than rely on appeals to authority. There were no shortcuts to Wisdom. Faust abandoned study and debate and turned to magic. Finally, the Biblical sages did not separate religion and ethics. While their specific beliefs were indistinguishable from those widely shared in Israel, their ethics were characterized by "a conviction that men and women possess the means of securing their own well-being" (p. 24). Divine intervention was not expected. The result was a pragmatic ethic which maintained that "virtue is its own reward" (p. 24). Contrast this with Faust who sacrificed his eternal soul for immediate gratification.

There is one point, according to Crenshaw, shared by the sages of ancient Israel and Faust (at least in Goethe's version)—the conflict of skepticism and faith (p. 191). The Wisdom literature brings skepticism into the Bible, emphasizing "self-knowledge through interrogation" (p. 193). Skepticism is not the same as unbelief, pessimism or cynicism. Skepticism involves doubt and questioning the "disparity between the actual state of affairs and a vision of what should be" (p. 191). In Goethe, it is this conflict which causes Faust to finally lose his wager with Mephistopheles but to be redeemed. In the Faustbook and Marlowe, however, Faust became world-weary and cynical, believing that it was futile for him to repent, and thus went to his damnation.

Faust and Ben Sira are thus two metaphors for how we can practice technology. Faust exemplifies the equation of knowledge with power, the quest for knowledge without limit and mastery without restraint, and gratification of our desires through magic. But Faust also symbolizes the frustration of those desires through the unintended consequences of our power and the terrible price we may pay for it. Ben Sira, on the other hand, can be a symbol for the wise practice of technology, a practice characterized by the recognition of limits, by moderation and restraint, by faith that skeptical reason and debate are a superior way to the good life and that trust in the doctrines of experts or the magic of the black box are not. Faust is the way of technological mysticism; Ben Sira can open a way to a redemptive technology.

Part II

REDEMPTIVE TECHNOLOGY

Chapter Six

Two Philosophers and a Metallurgist

> We can hold in our minds the enormous benefits of technological society, but we cannot so easily hold the ways it may have deprived us, because technique is ourselves....Technique comes forth from and is sustained in our vision of ourselves as creative freedom, making ourselves, and conquering the chances of an indifferent world.
>
> George Grant
> *Technology and Empire*

> Faith is shaped by experience, slowly but inevitably;...experience is likewise shaped by faith. We experience the universe about us, social and physical, differently because of the metaphors and models that focus and guide our perceptions. But we are not locked forever into religious or scientific preconceptions. The universe about us—valuational and sensory—presses back upon our metaphors and models. In this mutual pressure is found the ultimate dialogue that moves the history of thought and changes the face of civilizations.
>
> Frederick Ferré
> *Hellfire and Lightning Rods*

Technological mysticism, I said at the beginning, is belief in the universal efficacy of technology. Now we can see more clearly some of its manifestations. As an implicit religion, technological mysticism underlies much of our discourse. We have seen how it provides a template for some futurologists' utopian projections, grounds many people's identity, and forms the substrata of much of the media's reporting on computers. In the Faust myth it forms a central metaphor of advanced industrial society. At the core of its creed is the belief that knowledge equals power, and that power equals control. Of course, as with any creed, there are subtexts.

Power and control are always reserved for elites—scientists, engineers, industrialists—while its promises ring hollow for those whose lives are more controlled than controlling. There is also a gender subtext. Science and technology have too often been tacitly understood as being by men and for men, and as such the frames of many technologies have become part of the relations of ruling in advanced industrial society. In addition to being practice and identity, technological mysticism is also mystification. It makes technology into a black box, and when that happens technology becomes magic.

There are dangers in treating technology like magic. Scientists and engineers become priests and wizards whose pronouncements cannot be understood by the public and, implicitly, should not be challenged. This mystification applies as much to policy makers as it does to the general public, with the consequence that policy is habitually made on the basis of the grandiloquent visions of those who have something to gain, rather than upon any realistic assessment of the limits of technology. Such policies, from Aramis to Gigatext[1] to the Strategic Defence Initiative, repeatedly end up as financial and technical fiascos which waste resources and damage the natural and social environment.

As I said in the first chapter, the beginning of freedom from technological mysticism is demythologization. Unmasked, it begins to lose its power. But demythologization is only a first step. Jesus of Nazareth told the story (Luke 11: 24—26) of a man who had a demon cast out. The demon wandered about for a while and then returned to the man. There it found its former dwelling swept and tidied and empty. The demon moved back in, bringing with it seven others more evil than itself, and the man ended up worse off than before. In terms of the parable, demythologizing technological mysticism is like casting out the demon. Part I was, in Paul Ricoeur's (1970) terms, a hermeneutic of suspicion which used the tools of critical reason to unmask and demystify. But to stop there, to rely only upon formal reason, would—like the man left tidy and empty—invite the demon to return. Since we cannot live without myths, we must find new myths to put in place of technological mysticism. Since technological mysticism is an implicit religion, the best way to replace it is not another hollow attempt at imposing "reason" and "objectivity" but to *explicitly* embrace religious symbol and ritual in our understanding of technology. Remythologization is the other side of demythologization.

This may sound paradoxical. Is remythologization possible? How can one have a conscious or rationally founded and accessible mythology? Doesn't rational awareness automatically lead to demythologization?

---

[1] Gigatext is a notorious case in Canada. In the late 1980s the province of Saskatchewan invested millions of dollars in a scheme to use computers to translate all its laws into French. Not a single statute was ever translated.

Perhaps the way through the paradox is to recall the nature of symbols discussed in the previous chapter. Symbols resist being categorized as either "objective" or "subjective" because they are relational. By their very nature they merge the subjectivity of the interpreter with the objectivity of the signifier into a relational whole. Every symbol is therefore in some sense transcendent, "going beyond" both the subjective limitations of the interpreter and the objective limitations of the sign. As symbolic narratives, myths are also neither objective nor subjective; neither rational nor irrational. When Marlowe and Goethe, for instance, give us different versions of Faust, we do not argue whether one is "true" and the other "false" but rather accept both as profound interpretations of the myth. That both were writers consciously retelling the story does not make them any less meaningful. So demythologization and remythologization are not opposites but different moments in the hermeneutical circle, the ongoing process of interpretation. As Paul Ricoeur reflects: "Hermeneutics seems to me to be animated by this double motivation: willingness to suspect, willingness to listen; vow of rigour, vow of obedience. In our time we have not finished doing away with *idols* and we have barely begun to listen to *symbols*" (1970: 27). The circle of interpretation never stops. As Ricoeur concludes: "'Symbols give rise to thought,' but they are also the birth of idols. That is why the critique of idols remains the condition of the conquest of symbols" (1970: 543).

So unless we can replace technological mysticism with something better, simply unmasking it is futile and perhaps counterproductive. We need to tell other stories to take its place. We have to discover theological and ethical alternatives. Borrowing a phrase from Ursula Franklin (1990), I call Part II "Redemptive Technology."[2]

Redemptive technology is a symbol for a different way of practicing technology. It is not a case of choosing to develop this technology over that, although many of today's technologies could never meet its criteria. It is not another technical fix. Nor does it claim that a change in the social system will by itself solve technological problems.[3] While it is a different way of thinking about technology, it recognizes that different thoughts alone do not bring change. Redemptive technology requires a

---

[2] Others, of course, have undertaken this task as well. For two particularly insightful examples, see J. Mark Thomas (1987) and Ian Barbour (1993).

[3] I am reminded here of a comment made many years ago by the Yugoslav philosopher, Gajo Petrović:

> "Technology," it is claimed, is a mere instrument, which can be dangerous under capitalism, but which becomes "obedient" under socialism. There is something to this. *But the atomic bomb will not start producing edible mushrooms the moment we affix a socialist label to it.* (1967: 134, italics in original)

change in practice—in culture *and* organizations *and* techniques. The key to escaping technological mysticism is to see technology as a seamless web of all three of these aspects.

In this chapter I want to commence the process of building a redemptive technology practice by beginning to develop some of its cultural aspects, and will look at organizational and technical aspects in the next chapter. I concluded my critique of technological mysticism by looking at the Faust myth as a central metaphor of the modern world, so perhaps the best place to begin is with the philosophy of modernity.

The criticism of modernity (under a variety of names) has been a central feature of social theory for the past two hundred years. Most of these theorists agree that somewhere between five hundred and two-hundred-fifty years ago (depending upon what they see as the crucial events) a new type of society appeared in Western Europe and spread around the globe. Most also agree that this society is characterized by a shift from the organic world of magic to the mechanical world of science and the rising market economy (cf. Merchant, 1980; Saul, 1995). The debate has been over the causes and consequences of modernity. To Emile Durkheim, the essence of the modern world was the intensification of the division of labour. For Max Weber, modernity meant a process of rationalization which grew out of the Protestant Reformation. Karl Marx saw modern culture as the by-product of the rise of capitalism. To Herbert Spencer the modern world was progressive and peaceful, to Lenin it was characterized by war and imperialism. I would like to approach the question by contrasting two philosophers who, more than most, recognize the interplay of cultural, organizational, and technical aspects: George Grant and Frederick Ferré. Both are Christians (and address their arguments to a Western, Judeo-Christian audience), both make technology the centerpiece of their philosophy, and both emphasize the religious dimension of technology. But philosophy is not enough. I conclude the chapter with the practical insights of metallurgist Ursula Franklin. Confronting their thought will help develop the cultural aspect of a redemptive technology.

George Grant

The Canadian philosopher George Grant was a trenchant critic of modern society. A classical conservative,[4] he kept returning to the founding traditions of Western society in Greek philosophy and biblical religion as a standard against which to measure the modern world of liberalism and capitalism, of science and technology—and find them wanting. Grant

---

[4] Not to be confused with the ideology of the radical right which only calls itself "conservative."

espoused technological determinism as a consciously chosen and defended philosophical position. I will briefly explain his definition of technology and understanding of modern society before looking at his defence of technological determinism in detail.

Grant takes his definition of technology straight from Jacques Ellul's definition of technique: "the totality of methods rationally arrived at and having absolute efficiency (for a given stage of development) in every field of human activity" (Grant, 1969: 113; Ellul, 1964: xxv). This definition is much broader than the equation of "technology" with machines (which does open it, in Grant's work, to incorporate organizational and, especially, cultural aspects of technology practice), but it also is an essentialist definition which understands technology as autonomous and a determining force. It is essentialist as it defines technology as the embodiment of an instrumental reason which finds the "one right way" (conceived of as efficiency) in which technology will and must develop. Any given technology is as it is because of its own internal logic. Since a technology can only develop in one manner, it is both independent (that is, it cannot be developed differently than it is) and it is a determining force in society, for modern society *is* technology.

Technology is the essence of modern society, according to Grant, because "technique is ourselves" (1969: 137). We are technological beings because we are free, and we use that freedom to create in the midst of an indifferent nature. In other words, we have realized Francis Bacon's dream to master nature and thereby overcome blind fate. In doing so, however, we have fundamentally reshaped both society and ourselves. "Western technical achievement has shaped a different civilization from any previous," he says, because, "it moulds us in what we are, not only at the heart of our animality in the propagation and continuance of our species, but in our actions and thoughts and imaginings. Its pursuit has become our dominant activity and that dominance fashions both the public and private realms" (1969: 15). Where previous civilizations were organically linked to nature, for us nature is simply an object, something to be mastered and used as our freedom dictates. Where previous societies were wrapped in taboos derived from an understanding of human nature (as in medieval scholastic theories of natural law), moderns reject any conception of human nature except freedom (1986: 52). We create ourselves as much as we create the world.

It may at first seem paradoxical to speak of freedom while at the same time saying that technology is a determining force, but for Grant this is not the case. "Technique comes forth from and is sustained in our vision of ourselves as creative freedom, making ourselves, and conquering the chances of an indifferent world," he says (1969: 137). Technique is inherent in our understanding of freedom. Both the Bible and the Greek philosophers circumscribed human freedom with moral injunctions to love the other and seek the greatest good. The modern understanding of freedom rejects both—our freedom must be contentless if we are to be truly creative and to be able to master fate. "In myth, philosophy, and revelation," Grant declares, "orders were proclaimed in terms of which freedom was measured and defined. As freedom is the highest term in the modern language, it can no longer be so enfolded. There is therefore no possibility of answering the question: freedom for what purposes?" (1969: 138). If our freedom has no purpose external to itself, it must of necessity become technique. By understanding the world only as object, something apart from ourselves (1986: 36), we create the necessity for a method by which to know and control it. Our reason becomes instrumental, an account of means rather than of ends. To be free in the modern world is therefore to practice technology.

For Grant, modernity has consequences both for culture and for the organization of society. Modernity, he argues, is a culture of nihilism. True, he would say, we have a higher standard of living, better health and longer life, infinitely more commodities. Today we are more comfortable, but are we any happier?

> It is difficult to think whether we are deprived of anything essential to our happiness, just because the coming to be of the technological society has stripped us above all of the very systems of meaning which disclosed the highest purposes of man, and in terms of which, therefore, we could judge whether the absence of something was in fact a deprival. Our vision of ourselves as freedom in an indifferent world could only have arisen in so far as we had analysed to disintegration those systems of meaning, given in myth, philosophy and revelation, which held sway over our progenitors. For those systems of meaning all mitigated both our freedom and the indifference of the world, and in so doing put limits of one kind or another on our interference with chance and the possibilities of its conquest. (1969: 137)

We may have freedom. We may have the power and control that technology gives us. But we do not have a system of meaning (1969: 138).

We have lost, he claims, an understanding of the common good and of justice. In the modern world, an understanding of the good as that which all have in common has been transvalued into whatever is in the free individual's interest. Indeed, "the good" has largely been replaced in our discourse by the word "values," a term borrowed from economics and singularly devoid of substantive content (1986: 41). A "value" is whatever the individual chooses. Similarly, "justice" ceased to be that which was due to another just because they were human (e.g., the biblical injunction to protect widows and orphans) and became whatever was decided upon in a freely accepted contract between individuals (1986: 58-59). Justice has become no more than due process. At their best, modern theories of justice assert the equality of all individuals, but even at their best justice is only "something human beings make and impose for human convenience" (1986: 60). At their worst, "there is nothing intrinsic in all others that puts any given limit on what we may do to them...there is nothing belonging to all human beings which need limit the building of the future" (1986: 94-95).

This means, to Grant, that all technological societies live within a "common horizon" (1969: 30) which defines the parameters of their discourse. To step outside those parameters is to step outside the bounds of acceptable discourse. In modern society both the "right" and the "left" are founded on the twin premises of freedom and technology (however much they may quibble over the details). "When men are committed to technology," Grant says, "they are also committed to continual change in institutions and customs. Freedom must be the first political principle—the freedom to change any order that stands in the way of technological advance" (1965: 72-73). A capitalist democracy is a contractual society of free and equal individuals which has, in principle, no place for distinctions of race or gender or class (1969: 87-89), or indeed of any kind of distinction based on identity, community, or local culture (1965: 54). We have created, in his words, the universal and homogeneous state. In the modern world, a world in which everyone is the same and everyone is alone, there is no place for the Other.[5]

This is the context for Grant's defence of technological determinism. I would like to look in detail at the argument he presents in his essay "Thinking About Technology" in *Technology and Justice* (1986). Grant begins his essay by reiterating his belief that the modern world differs from all previous societies in that it has a unique account of the relationship between knowing and making. This fundamentally changed what we mean by "art" and "science," so while there are continuities of language (e.g., "technology" and "technique" are derived from the Greek word *techne*[6]), our understanding of what is meant is profoundly different from that of the ancients.

The main portion of the essay is a refutation of the statement: "The computer does not impose on us the ways it should be used." To begin with, Grant argues, computers are not neutral instrumentalities, abstracted from "the events which have made possible their existence" (p. 21). Computers are built within a paradigm of knowledge which is central to modern

---

5 Grant makes this the central theme of his political philosophy, which laments the impossibility of community, and the dire implications this has for the continued existence of both English and French Canada.

6 Interestingly, our words "art," "artist," and "artisan" are derived from the Latin words *ars* and *artis*, which were Latin translations of the Greek *techne*.

civilization—it is taught in the schools and implemented in office and factory. They are integral to our *civilizational destiny*, the "fundamental presuppositions that the majority of human beings inherit in a civilization, and which are so taken for granted as the way things are that they are given an almost absolute status" (p. 22). It is "the destiny which *does* 'impose' itself upon us, and therefore the computer does impose" (p. 23). While it is this paradigm which "imposes" upon us, not this or that machine, the machines cannot be abstracted out from the paradigm of knowledge which created them.

Further, he says, how do we understand "the ways" that the computer supposedly does not impose on us? He makes three arguments. First, computers increase the tempo of the homogenizing process already inherent in the modern project. Computers are part of information technology, and the very "word 'information' is itself perfectly attuned to the account of knowledge which is homogenizing in its very nature"(p. 24). Grant compares the effects of computers today with that of automobiles seventy-five years ago. Who in the first years of this century could imagine the superhighways, parking lots, pollution, and congestion "imposed" on us by the necessities of the automobile?

Second, Grant points out that computers can only exist in societies which have large corporate institutions in which to design, finance, build, sell, and use them. Therefore these necessary preconditions for computers exclude some types of community and permit others (p. 26). Whether the society calls itself capitalist or socialist, computers require the same kind of large organizations and therefore impose a particular form of social structure.

Third, Grant states, if technology is neutral, then how it is used will be determined by the society's dominant political philosophy. But the same understanding of reason produced both technology and modern political philosophy. "The ways that computers have been and will be used cannot be detached from modern conceptions of justice," he says, and "these conceptions of justice come forth from the very account of reasoning which led to the building of computers" (p. 27). Computers therefore cannot be neutral instrumentalities.

Finally, he looks at the nature of the *should* in "should be used." He recalls that the essence of the modern world is a contentless liberation which makes, as its consequence, everything provisional and subject to change. To say "'the computer does not impose on us the ways it *should* be used' asserts the essence of the modern view, which is that human ability freely determines what happens. It then puts that freedom in the service of the very 'should' which that same modern novelty has made provisional" (p. 31). So by not imposing how they should be used, computers impose a particular understanding of *should* from which we cannot escape in the modern world. Its account of freedom is the *only* option.

When we chose a technology, Grant concludes, we like to think of ourselves in a supermarket full of options (p. 32), but in reality, we buy a "package deal." Coming with any technology is the whole package of paradigms of knowledge, political philosophies, forms of social organization, and accounts of reason, justice, and what is good. "Technology is the ontology of the age," he declares (p. 32). We are determined by technology because technology is the essence of the modern world.

In conclusion, Grant sees modernity as our fate. Unlike Romantics, he does not dream nostalgically for return to a bygone era or a lost golden age—the past is dead and cannot be resurrected, even were it desirable to do so. He gives no indication that it would. The modern world has made real gains over the brutalities of previous societies (1986: 59-60). Yet the price we have paid for our freedom in technological society is both very high and inescapable. We may have power and a high standard of living but we have lost those traditions which gave our lives meaning and which provided standards of justice and the good through which we could measure and judge our communities and ourselves. We no longer have a way, to use an old Quaker saying, of speaking truth to power. He has no illusions about the possibility of restoring tradition in a technological society, but without some transcendent understandings of justice and the good we are trapped in the meaninglessness of the modern world. Grant offers no program. He can only lament that which has been lost.

Determinism and Modernity

George Grant issues a powerful indictment of modern society. Analyzing his philosophy is not easy, however, because there is a paradox involved in criticizing Grant from a constructivist standpoint. Constructivism first appeared as an account of the development of technology in opposition to the doctrines of technological determinism and, as such, placed its central emphasis on human freedom (*agency* in the jargon of sociology). To criticize Grant from a constructivist standpoint is therefore to accept his analysis of the essence of modernity as freedom. This paradox is inescapable because we *are* modern. All criticisms of modernity—including Grant's—are criticisms from within.

There are two issues in Grant's thought which are crucial to the project of redemptive technology. I want to affirm much of his analysis of the devaluing of tradition and the effect this has had, but reject his theory of technological determinism. I will look at technological determinism first.

George Grant's defence of technological determinism is, to put it bluntly, wrong. Grant's definition of technology rests on unsupported assumptions (a weakness he borrowed from Ellul along with his definition) which do not stand up to examination. There are several logical and empirical problems.

First, Grant offers no account of the development of technology. He starts from what Bruno Latour (1987) calls "ready made" technology, that is, machines and techniques which have already completed a long process of development. Thus he talks about the effects which computers impose upon us, but does not say how computers came to be. If we do look at the actual development of technology, its definition as "the totality of methods rationally arrived at and having absolute efficiency" becomes problematical, as do claims of a controlling "paradigm of knowledge" (1986: 21-22). Tracy Kidder's (1981) story of the development of a new computer, Wiebe Bijker's (1995) history of fluorescent light bulbs, or the tale of Aramis (Latour, 1996), the personal rapid transit system for Paris recounted in the previous chapter, give us a very different picture. Building the Eclipse MV/8000 computer was not what one could call a textbook case of applied science or management. It was born out of internecine politics at Data General Corporation and its rivalry with DEC, its engineers trained for the project by playing computer games, project management was by "the seat of the pants," and its development was a frustrating process of trial and error. In other words, it was altogether typical of the computer industry. Far from displaying a drive to "absolute efficiency," fluorescent light bulbs actually became less efficient as General Electric and the utility companies battled over their effect on power consumption (and utility profits). Aramis was inspired by visionaries and executed by bureaucrats. Over seventeen years the project squandered half a billion francs without doing significant research into the central problems the technology would have to surmount. In sum, the development of technology is a contingent as well as constrained process, often chaotic, and usually immersed in politics. Its methods are trial and error as much as "rationally arrived at." Efficiency is politically defined. When we look at what actually goes on in the lab, the "paradigm of knowledge" more often appears after the fact, as a justification for the discovery or invention rather than its cause (Collins, 1985; Latour, 1987). By failing to become concrete, Grant lacks what constructivists call *symmetry* and *interpretive flexibility*, that is, he treats what should be the *explandum* (that which needs to be explained—in this case, the form of a particular technology) as the *explanans* (the cause or reason). In other words, he *assumes* what he needs to *prove*.

Now, Grant might respond, philosophy does not pretend to be sociology and he did not intend to give a detailed account of the development of computers or any other technology. But either the "paradigm of knowledge," "civilizational destiny," or "account of reason" (1986: 27) directly shapes the development of technology, in which case he would have to give a detailed account showing how it does so (and why the empirical reports of technological development cited above are wrong), or it is a worldview which determines *every* event *by definition.* His technological determinism is a "Whig history" which sees the present as inevitable. This is no more than *post hoc* reasoning, a rationale made up after the fact to explain whatever happened.

The same problems accrue to Grant's final argument. What does he mean when he says computers are a "package deal" which "enfolds us in its own conceptions of instrumentality, neutrality, and purposiveness...technology is the ontology of the age" (1986: 32)? If he is only saying that Bacon's vision has become dominant in the modern age and that this vision is a prerequisite for the existence of computers, his statement is true but trivial. However, he seems to be saying more, that computers are a necessary outcome of modernism. The structure of his argument is that given the modern worldview ("civilizational destiny," paradigm of knowledge), then computers are a necessary outcome, therefore computers impose upon us because the modern worldview imposes upon us. But this argument is based on a logical fallacy. The existence of a worldview may be a necessary part of an explanation, but it is not sufficient to account for any particular happening. He would have to become empirical to show how and why this account (and not another) explains the development of computers, which, as we have just seen, he does not do. In sociological terms, he does not account for the relationship between *agency* and *structure,* that is, between free individuals and the often constraining groups and organizations of which they are a part. As soon as theory becomes historically concrete, however, it begins to become contingent. We see that transit systems or light bulbs or computers are not the inevitable product of the logic of modernity but the outcome of considerable work and struggle by individuals and groups. The result easily could have been different. Had more resources been put into research rather than production, perhaps by today Aramis would have been a model transportation system (Latour, 1996). Had the backroom politics between GE and the utility companies gone differently, perhaps today fluorescent lights would be known for their low power consumption rather than their high intensity (Bijker, 1995). Had Tom West not been so doggedly stubborn in overcoming higher management indifference and keeping his engineering team going, the project to build the Eclipse MV/8000 computer could easily have failed (Kidder, 1981). Only the *assumption* of technological determinism can lead from general preconditions (the modern scientific worldview) to specific results (the current form and use of computers), but this is what Grant's essay set out to *prove*. He ends up begging the question. His reasons and his conclusion are the same.

While Grant's understanding of technology is gravely flawed, we should not overlook the importance of his insights into the nature of modernity. There are some flaws here too. Grant offers an abstract, single-factor theory which has trouble trying to explicate the complexities of empirical history, but this should not distract us from the significance of his insights.

Chief among them is the inability of tradition to provide ends or goals for the modern world. Whatever the cause, tradition can no longer provide a standard of judgment, a definition of the good, or a source for justice. It is not that these ideas have disappeared (as Grant's own writings testify) but rather that they have ceased to be foundational for our culture. They have become only one weak voice lost in the babble of conflicting doctrines and ideologies, one more option for individuals to choose from. Others have pointed this out as well, but they usually argue for a return to the traditions of the past. Grant dispels such wishful thinking. We cannot wish away the past few centuries of life in the West (as much as some Romantics and Fundamentalists may try). The significance of Grant's insight is that he names what is absent. Where modern society devalues justice and the good to no more than another option among many, Grant insists on their centrality for life in community. Without them community disappears and all are absorbed into the universal homogeneous state.

Here the paradox with which I began this commentary returns. To the extent that constructivists destroy theories of determinism and inevitability, they reinforce precisely that contingency which Grant places at the heart of modernity. We can only criticize the modern from within, for we no longer have any foundations, and we must be aware of the risks and snares when we do. But this paradox also opens possibilities. To strip Grant's ideas of technological determinism is to strip them of some of their pessimism as well. The existence of technological mysticism (and other forms of implicit religion) shows that people are perhaps not quite as modern as many theorists think. Myth and religion can be found at the heart of what Grant sees as the very essence of modernity. Far from being a total system, modernity itself is only one story among others, albeit the one told the most loudly and frequently. By reducing everything to an option, including itself, modernity opens the door to going beyond the modern. A philosopher who tries to guide us through that door is Frederick Ferré.

Frederick Ferré

At first reading, the contrasts between Grant and Ferré stand out more than the similarities. Grant laments the past, Ferré looks to the future. Where Grant uses the traditions of Greek philosophy and Biblical religion to criticize the modern, Ferré tries to go beyond to the postmodern. Grant is pessimistic, Ferré is cautiously optimistic.[7] But both are needed for the project of a redemptive technology.

Ferré argues that we are living at the end of the modern and (if humanity is to survive) the beginning of a postmodern age. He summarizes his thesis:

> Modern science, besides being a secular set of particular practices designed to provide specific answers to limited questions, became the generator of a religious movement that vanquished one civilization, the medieval, and gave birth to a new world in its own image. The crisis of the passing away in our time of this modern world so generated is, therefore, the crisis posed by modern science being extended beyond inquiry to theopany, the crisis of scientism as mythic matrix for our lives. The nub of the search for a livable postmodern world is the task of finding a worthy successor for scientific consciousness without abandoning the genuine virtues of science itself.
> (1993: 8)

For him, modernity is defined by the consciousness of the industrialized "North," a consciousness decisively shaped by science. Flaws in the modern consciousness are destroying the environment and will inevitably bring about the end of modernity itself. Religion, in the form of a postmodern faith, is the key to creating a new *mythos*.

He begins with a (perhaps excessively) functionalist definition of religion as that which is centered on primary values and in which feelings, perceptions, beliefs, and behaviours are more important than doctrines. Modern science, he says, is more than a methodology for describing empirical reality. It is fundamentally religious. "Scientific practice," he says, "involves both sacred ritual and religiously sanctioned ethics" (p. 12). Science is both grounded in primary values and extends beyond empirical observation to form a metaphorical way of understanding the world, what he calls a religious world model. *Scientism* is the religious world model of the modern consciousness.

The problem with scientism is that its obsession with quantity and "objective" consciousness alienates us from concerns for quality, its reductionism alienates us from perception of the whole, and its fixation upon technique alienates us from humanistic visions of the universe. This

[7] Although, as I remind my students when they complain that some theorists are "too pessimistic," optimism and pessimism are not categories of truth or falsehood.

alienation has as its direct consequence the environmental crisis. Because nature is understood only as "stuff," an object to be manipulated, the modern consciousness recognizes neither limits to, nor responsibility for, exploitation. The result is the threat of nuclear war and exponential growth in both population and the consumption of resources. Since exponential growth cannot continue indefinitely on a finite planet, the modern world, and perhaps humanity itself, will, sooner or later, *inevitably* come to an end.

While Ferré is sharply critical of scientism and what he calls *technolatry* (which is virtually synonymous with technological mysticism), this is not another Romanticist diatribe. He appreciates the insights of empirical science and understands that human beings are dependent upon technology. What is needed, he says, is to liberate science and technology from scientism and technolatry and to develop a new paradigm built upon creativity, limits, and holism. He suggests that ecology provides a model for such a postmodern science.

For Ferré, ecology stands out as a model of postmodern science for three reasons (pp. 93-96). First, it provides a way of thinking which is completely analytical and rational, yet different from reductionist and mechanical models of science. Second, ecology engenders a different vision of reality, one which portrays the world "as an endlessly complex network of organic and inorganic systems locked in constant interaction" (p. 94). An ecosystem cannot be reduced to a single underlying factor but must be seen as a mutually interrelated whole. And third, in its emphasis upon equilibrium and limits, ecology involves a set of values different from those of control and mastery associated with scientism. Ecology is important for Ferré because it offers a model of a creative yet rational holism which differs sharply from the mystical holism of Romantic and New Age movements.

Ferré is just as critical of those religious groups which attempt to substitute religious doctrine for scientific findings. Ferré takes the title of his book, *Hellfire and Lightning Rods*, from a sermon his father remembered hearing as a youth, in which the preacher denounced farmers who showed their lack of trust in the Lord by installing lightning rods on their barns. The title becomes a metaphor for the futile way religion has tried to confront science and technology.

In spite of this, he argues that the path to a postmodern consciousness lies through religion. Much of his book is an examination of myths, both those of modernism and those which offer potential for liberation. While he recognizes some value in premodern science and New Age mysticism (such as shamanism, Wicca, the occult, etc.)[8] which raise many of the themes, such as creativity and holism, that Ferré identifies as crucial, he nevertheless does not see much hope in them. Like Grant, he argues that we simply cannot turn back the clock. "Modern society is too complex, and there are too many people to expect that postmodern society will be able to live by shaking acorns from the trees and picking up small round stones for its fires," he says, "the postmodern world will require *post*modern technologies, not *pre*modern ones" (p. 147). Furthermore, there is a serious danger that these Romantic movements will throw out the truths of modern science along with scientism, nor are very many of them sufficiently self-critical to forestall exploitation arising from within their own ranks. "The dreadful specter arises of a postmodern society combining the worst aspects of exploitative modern consciousness with the worst aspects of rampant credulity and superstition," he concludes (p. 148). The ease with which New Age techniques have been appropriated by large corporations to enhance sales and employee productivity (Ziguras, 1997) would seem to bear out these apprehensions.

The only traditions, Ferré argues, which have the potential to move the whole society through the period of transition between modern and postmodern societies are Judaism and Christianity because only the "mainstream" religions are rooted deeply enough and still command enough adherents to become viable possibilities.[9] Even so, they will have to undergo major revisions. The Christian worldview, "fashioned from Biblical images perceived through Aristotelian categories" (p. 151), has been chronically unable to value nature for itself. Nature may be valued as the creation of God, particularly if the Biblical injunction to "have dominion" is understood as stewardship rather than as a license to exploit, but either way the value of nature is secondary and derived. Consequently, "restraint has not been a noteworthy effect of mainstream Christianity in the past" (p. 153). Nor is it likely that mainstream Christianity will be able to retain appeal for its still largely premodern images of God. Furthermore, Christianity has just as many credibility problems as do New Age movements. Ferré concludes that "the mainstream version of the Christian worldview is a dubious bridge to postmodernity" (p. 155). However, "might it be that Christianity could rise to the historical challenge once again and develop a *non-traditional but still Christian* version qualified to carry the new day as once it carried the old?" (p. 155). Ferré believes it can, and calls for a *multimythic organicism* derived from Alfred North

[8] For a more complete discussion of New Age movements, see Hanegraff, 1997; Heelas, 1996.

[9] Ferré is aiming his argument at a North American and European audience. He does not address the potential of religious traditions indigenous to other peoples, although his multimythic organicism would seem to be open to including their symbols.

Whitehead's process philosophy[10] as the best hope for developing a postmodern faith.

Multimythic organicism could provide a bridge between the alienation of modern consciousness and whatever will develop in the future, Ferré contends. He defines it as: "A religious stance that affirms as legitimate and exciting the possibility of pluralism in mythic imagery within a context of undergirding fundamental values. It is not a religious response without organizing imagery but, rather, one with many value-focusing sets of myths welcome within it" (p. 168). Instead of mysticism or a futile attempt to retain the past, it builds upon the pluralism and fragmentation of modern society and, using ecology as a model, tries to restore an organic relation to nature and community. It is characterized by the fundamental values of creativity, limits, and holism. Creativity arises out of the "balance between necessity and spontaneity" (p. 168). It recognizes the importance of adventure and growth to the human spirit. In the coming world of scarcity, brought about by the depletion of the planet's resources, creativity will be necessary to find appropriate technologies to enable us to do more with less. "One of life's proper responses to shortage is to create adequate supplies in responsible ways," he says, "Christian imagery, not only reflecting on the original creative God but also on Christ's multiplying the loaves and fishes or bringing wine from water, will amply model this truth for a world that may need encouragement to avoid the self-defeat of despair" (p. 199). Creativity may be dangerous without limits, however. A postmodern faith will abjure runaway creativity in favour of balance between growth and death which recognizes proper scale and the limits to growth. Finally, tying creativity and limits together is a sense of holism. While his model of holism is grounded in ecology, here too Christian imagery may help establish "the balance between local differentiation of function and holistic mutuality of connection" (p. 168). By making connections between the local and global, holism "may be able to lay the basis for fair play and the dignity of widely shared participatory engagement" (p. 200) which is the antithesis of Grant's universal homogeneous state.

For Ferré, the important thing is to begin with wherever we are in the modern world and move towards a postmodern faith. A new consciousness can only merge out of dialogue, it cannot be imposed.[11] "We cannot be mythic engineers," he says (p. 196). But out of dialogue emerges the possibility of hope.

---

[10] Process philosophy is complex but has played a central role in the dialogue between religion and science. Its central idea is that all things, including God, are interrelated and in process and unfolding. See Whitehead, 1929, 1933; Barbour, 1990.

[11] Ferré does not specify who participates in the dialogue, or in what forum it should take place. I will try to address these questions in the next chapter.

Consciousness and Modernity

For all their differences, by now some of the similarities between George Grant and Frederick Ferré should begin to appear, as well as places where they correct each other. Both concentrate on the cultural aspect of technology practice. While both speak of making connections and the importance of seeing the whole (and do so more than many others) neither pays enough attention to the technical and organizational aspects. Both would benefit from an even stronger sense of the seamless web of technology practice.

Both are philosophers of consciousness (cf. Habermas, 1991), Ferré even more than Grant. To Ferré, modernism is consciousness derived from science and the extension of science into scientism and technolatry. Some of the same criticisms leveled against Grant apply to him as well. Like Grant, Ferré does not concretely demonstrate the connections between a generalized consciousness and specific events. Science may be necessary to an explanation of modernism, but is it sufficient? While in the passage used as an epigraph to this chapter he mentions how nature "presses back upon our metaphors and models" (1993: 195), notably absent from his account is any investigation of social structure and, in particular, of capitalism and industrialization. Is science responsible for the destruction due to an economic system built upon the need for ever-increasing consumption of resources? Is alienation in society caused by scientistic reductionism or by the reification that occurs when every aspect of life becomes a commodity in the marketplace? Can liberation occur by developing a new consciousness, or must we also address questions of power? These questions point to the complexity and interrelatedness of our situation, especially the importance of institutional, and particularly economic, factors. Concentration on cultural aspects of technology practice alone is just as short-sighted as is a technical fix. The scientific consciousness is only one part of modernity (albeit a central one). A redemptive technology will have to address institutions as well.

Ferré very clearly argues from within modernity, and quickly dismisses both Romantics and Fundamentalists who dream of a simpler age. The modern emphases on liberation and creativity are central to his thought. This gives him both optimism and direction (where Grant can only lament), but it does create some blind spots. While Ferré has hope that revised Jewish and Christian traditions may constitute the basis for a multimythic organicism, he pays insufficient attention to those elements of the tradition which are central to Grant—the Biblical conception of justice and the Greek ideal of the good. Without them his "fundamental values" lack orientation. Creativity is wonderful, but what is to give it direction? Liberation is essential, but liberation to what? If the only limits are those imposed by nature, what restricts what we can do to other people? Multimythic organicism is nebulous. Benito Mussolini and George Sorel called for organicism earlier in this century, but I am sure they do not provide a model for what Ferré has in mind. He is critical of New Age and other Romantic movements, but does not adequately distinguish their mystical organicism from his own. Part of his problem is that "post-modernism" is too vague a concept. Like "post-" anything, it is defined only in terms of what it is not. Ferré is quite right in rejecting a rigid or static utopia but a more clearly articulated vision, itself grounded in commitment to justice and the good, could provide necessary direction for dialogue.

The virtue of Ferré's philosophy is that we can start from wherever we are at. We can begin to change without having to complete a whole new mythos first. While still needing further definition, the merit of multimythic organicism is that it builds upon the fragmentary culture which modernism has left us and starts to reconstruct a new system of meaning. Since multimythic organicism is always open, in process, and unfolding, it escapes the rigidity of utopianism while its fundamental values avoid the directionless growth of the market. It provides us with some principles for a redemptive technology.

Ursula Franklin

Having explored the ideas which two philosophers might contribute to the cultural aspects of a redemptive technology, it is only fair to add the voice of the person who coined the phrase. Ursula Franklin studied experimental physics in Berlin and taught metallurgy at the University of Toronto. Her reflections, based on a lifetime of practical experience in science and technology, bring a concreteness to discussion of the culture of technology which Grant and Ferré sometimes lack.

Franklin (1990) sees technology as a practice. In today's world, she says, we have two types of technology practice: holistic technology and prescriptive technology. Each practice has several characteristics (see Table 6.1). First, holistic technology is work-related, that is, it is oriented to achieving specific tasks and making practices easier. By contrast, prescriptive technologies aim to control the process of work itself. So for example, a free-standing word processor is work-related in that it makes the task of creating a document easier (p. 18). But if word processors are linked together in a network which allows management to monitor and supervise the typists, then the technology is control-oriented.

Second, holistic technology embodies a vision of labour as craft. Its division of labour is specialization by product. The doer is in control of the work process as, for instance, a potter who specializes in various types of pots but still controls the work of making a pot from beginning to end (p. 19). On the other hand, prescriptive practices are characterized by extreme division of labour and specialization by process, rather than by product. An auto assembly line, where workers each perform a single, specialized, and repetitive task, is a good example. Instead of being doer (or user) oriented, prescriptive technologies are designs of compliance in which "a workforce becomes acculturated into a milieu in which external control and internal compliance are seen as normal and necessary. Eventually there is only one way of doing something" (p. 23).

Table 6.1

Franklin's Types of Technology Practice

HOLISTIC TECHNOLOGY	PRESCRIPTIVE TECHNOLOGY
Work-Related	Control-Related
Craft	Division of Labour
Specialization by Product	Specialization by Process
Doer (or User) Oriented	Designs of Compliance
Growth Model	Production Model

Finally, holistic technologies are characterized by a growth model, while prescriptive technologies are built upon a production model. With the metaphor of "growth" Franklin underlines the importance of *appropriate* size and scale (p. 26-27). Growth, she says, "can only be nurtured and encouraged by providing a suitable environment. Growth occurs; it is not made. Within a growth model, all that human intervention can do is to discover the best conditions for growth and then try to meet them" (p. 27). The production model is very different. The emphasis here is on mastery, prediction, and control. Every variable is manipulated to achieve the maximum output at the minimum cost. Where the growth model recognizes the links between industry and the larger social and environmental context, the production model treats such links as *externalities* which are irrelevant to the productive activity itself (p. 27).

Prescriptive technology is the dominant practice in advanced industrial society. This practice is built on the fragmentation of reality into spheres of expertise, with little reciprocity between them (p. 40). One must be an expert (or command sufficient organizational power) to be admitted to this practice. Scientific management (or "Taylorism") with its extreme division of labour and centralized control of the work process by management is the epitome of today's practice. It is a technology which is anti-people (p. 77).

Because the environment is regarded as an externality, industry frequently shows little concern for, or often even awareness of, the damage it does to nature. Franklin makes a distinction between divisible and indivisible costs and benefits (p. 69). An indivisible cost or benefit is one which is shared by everyone, whether they contributed to it or not. Divisible costs or benefits, on the other hand, are shared only by some. In a capitalist economy, industry usually operates on the basis of divisible benefits and indivisible costs. For example, an industry may be very profitable for its managers and shareholders (a divisible benefit) but treat nature as a "free good" by dumping its waste into the air and water (an indivisible cost). Only a few benefit from the corporate profits, but everyone pays the costs of environmental degradation. Clean air and water, on the other hand, are indivisible benefits since they accrue to everyone. For Franklin, the common good is made up of indivisible benefits (p. 70).

The alternative to continued domination by prescriptive technology practices, according to Franklin, is to build a redemptive technology. The most central principles of redemptive technology are reciprocity and holism. Reciprocity is fundamental to dialogue and is therefore the essence of democracy, as well as justice, fairness and equality. Holism is the key to recognizing limits, building community, including marginalized people, and protecting nature.

Franklin defines reciprocity as "some manner of interactive give and take, a genuine communication among interacting parties" (p. 48). Current technology practice does not emphasize reciprocity. The control syndrome structures technology practice to produce a one-way flow of information, from those who are in control to those who are being controlled. Television is an outstanding example. Television creates passive viewers whose only options are to watch what the broadcaster sends them, changing to another channel (and seeing what *its* broadcaster wants to send), or turning the set off altogether. While many technologies today include feedback, Franklin insists that this is not the same as reciprocity (p. 49). Feedback is a control mechanism which allows a technology to become more efficient but which does not change its overall dynamics. Reciprocity, on the other hand, could fundamentally alter a technology because it would necessarily entail a shift in structure and power. A redemptive technology would structure technology practice to emphasize mutuality and dialogue. It would make technology practice democratic.

Dialogue is necessary but is not in and of itself sufficient to build a redemptive technology. Franklin also insists on commitment to justice, fairness, and equality (p. 123). A practice may be democratic and still be unfair or unjust (which is why charters of rights in democratic countries put so much emphasis on protection of minorities). A redemptive technology practice will have to conduct its dialogue on a global scale to ensure justice, fairness and equality. It will have to become holistic.

Holism is the other central principle of redemptive technology. A redemptive technology is inclusive, incorporating workers and users into its dialogue, sharing power, and building community. Because it aims at mutuality and dialogue, rather than only at a virtuoso performance or economic gain, it can more easily recognize limits and better protect nature. In contrast to Ferré, however, Franklin's holism is far more concrete.

For Franklin, the key to a more holistic practice is to pay close attention to the organizational aspects and not have our attention focused on the artifacts alone. "What turns the promised liberation into enslavement," she says, "are not the products of technology *per se*—the car, the computer, or the sewing machine—but the structures and infrastructures that are put into place to facilitate the use of these products and to develop dependency on them" (pp. 101-102). Once we understand technology practice as a seamless web, concrete steps become possible to ensure holism. For example, one thing Franklin suggests is making a checklist a regular part of our public discourse about any project:

> Should one not ask of any public project or loan whether it: (1) promotes justice; (2) restores reciprocity; (3) confers divisible or indivisible benefits; (4) favours people over machines; (5) whether its strategy maximizes gain or minimizes disaster; (6) whether conservation is favoured over waste; and (7) whether the reversible is favoured over the irreversible? (p. 126)

Such a checklist would help demystify the project and clarify civic and technological issues for both the public and policy makers. Another suggestion Franklin makes for a more holistic technology practice is to require both the public and private sectors to keep three sets of books (p. 129). One set would be the dollars and cents accounts currently required, but "with a clear and discernible column for money saved" (p. 129). A second set of books would catalogue "the human and community gains and losses" (p. 129), such as the number of people employed or unemployed, the quality of work (deskilling, etc.), consequences for the community's "quality of life" (traffic, housing, tax base, etc.), and other social "externalities." The third set would record the environmental gains and loses. The strength of these recommendations is that they holistically incorporate all three aspects of technology practice. Furthermore, they would be no more difficult to do than the impact assessments currently required by law in many jurisdictions.[12]

Principles for a Redemptive Technology

Franklin, a metallurgist, proposes principles remarkably similar to those of the philosophers Grant and Ferré. Franklin, however, speaks much more concretely out of a lifetime immersion in the practice of technology. Together, all three give us a set of principles which can become a guide for the cultural aspects of a redemptive technology practice. In summary, these principles are:

- *Search for the common good.* We have to look for that good which the community has in common. This is not just the good of a majority of individuals as defined by the sum of their self-interest but a good that transcends self-interest. As Grant reminds us, this concept of the good was a basic part of the Western tradition until it was submerged by modernity, and Franklin's idea of the common good as indivisible benefits provides a useful way to begin reclaiming that part of our tradition. I say the principle is the *search* for the common good because it is not something that can be defined a priori or perhaps ever be fully realized, but it provides a goal and a standard by which we can evaluate technology practice.
- *Commitment to justice.* Without justice there is no community. As St. Augustine said long ago, without justice a society is no more than a gang of bandits. But justice has to be more than due process, or simply giving to someone what is their due based on merit. As in the Biblical and Islamic traditions, justice must give priority to the poor and marginal, to the "other."
- *Creativity.* While we build community, we also have to give proper scope to the growth and contribution of individuals. As Ferré insists, redemptive technology practice is neither "anti-technology" nor Romantic but seeks new and more appropriate technologies with which to build the future. This principle demystifies the modern notion that creativity is something mysterious and limited to a few artists and restores it to the world of work.
- *Respect for limits.* Size does matter. Redemptive technology rejects the Faustian quest for limitless power, wealth, and growth and seeks to build technologies of appropriate size and scale. Balance and respect for nature and human communities set the proper scope for creativity.

---

[12] For more on technology assessment see chapter seven.

- *Reciprocity.* Redemptive technology is not a utopia. It is the product of dialogue and debate among all members of the community. It rejects the cult of expertise and affirms participatory democracy. This means actual democratic decision making in both government and the workplace, and not just the outward show of "consultation."
- *Holism.* Holism means taking all three aspects of technology practice seriously. A redemptive technology practice is a seamless web of technical, organizational, and cultural aspects which rejects reductive and mechanical formulas or the notion that society and the environment are irrelevant externalities to technology. Redemptive technology begins to overcome the fragmentation of modern culture by inclusively uniting both local communities and global concerns. It aims to bring together both science and religion.

All six of these principles are needed to reinforce and correct each other. As principles for living in the real world, not a conflict-free utopia, there are sure to be friction between them. It is easy to imagine disputes arising between creativity and limits, for instance, or between reciprocity and justice (since democracy may in some circumstances be unjust). Such discord can be resolved, however, by reference to other principles. So, for example, when individual creativity comes up against the respect for limits, the conflict can be resolved by reference to reciprocity, holism, and the common good. Perhaps individuals will need to curb their ambitions, or perhaps a deeper understanding of what limits are actually required for maintaining community and balance with nature will emerge.

These principles can form the basis for a wise use of technology. In that they are balanced, limited, and practical they begin to embody the wisdom of Ben Sira into the cultural aspects of a redemptive technology practice. They also provide criteria for evaluation as we look towards the technical and organizational aspects.

Chapter Seven

Technology in the Good Society

> Our individualistic heritage taught us that there is no such thing as the common good but only the sum of individual goods. But in our complex, interdependent world, the sum of individual goods, organized only under the tyranny of the market, often produces a common bad that eventually erodes our personal satisfactions as well.
>
> Robert Bellah et al.
> *The Good Society*

What is to be done? The question is at least as old as social science, and it still bedevils anyone with an analysis of a social problem. Since technology is a practice we must become practical. While theology and philosophy address some of the cultural dimensions of technology practice, we must confront the technical and organizational aspects as well. We need to embody the principles of redemptive technology in institutions or they will remain abstract.

At the end of chapter five I contrasted Faust and Jesus Ben Sira as two symbols for technology practice, the first exemplifying technological mysticism and the second the way of Wisdom which opens possibilities for a redemptive technology. Wisdom is always practical, it is characterized by restraint and prudence. If we are to be wise in our development and use of technology we must develop some means of assessing it. Assessment involves judgment and evaluation. It is the practical application of criteria and standards to make a decision as to the merit, value, and appropriateness of whatever is being assessed. This means that values, ethics, and politics have to become an integral part of our thinking about technology.

The idea of assessing technology is not new, but most attempts to do it have been caught in a paradox: The more our knowledge supposedly gives us the power to control our social and natural environment, the more people feel that technology itself is either in control or out of control (which amounts to the same thing). Technology assessment (TA) was a branch of policy studies first developed in the United States in the late 1960s which aimed to enhance the benefits of new technological development and to avoid unforeseen or adverse social and environmental impacts. It was an attempt to regain control over technology through knowledge. It failed. Examination of why it failed may prove instructive to developing a wiser program for assessing technology.

This chapter will begin by examining the origins, assumptions and contradictions of technology assessment. I will conclude by looking at some ways we can go beyond technology assessment to a redemptive technology practice.

The Ogburn Legacy

The American sociologist William Fielding Ogburn (1886-1959) did more than most to entrench the metaphor "social impact" in North American discourse. Ogburn and his associates (known as the "Ogburn School") flourished from the late 1920s to the late 1950s but their way of thinking continued and became the basis for technology assessment. Many of their ideas still dominate our discourse. Their chief concept became known as *cultural lag theory*. Its central thesis was that technology induces changes into society more rapidly than the culture and other institutions can adjust to them, creating a "lag." The Ogburn School made four arguments to support this idea.

First, they argued that technology causes social change. As Ogburn put it, "In order of time, the invention comes first and the changes in society follow, and secondly...these changes frequently require time before they develop into real readjustments to the demands of the invention" (1946: 5). In saying "demands of the invention," Ogburn implied that there is a certain "logic" inherent in how a machine is used. Technology may be employed in a number of ways, but each choice involves *inevitable* consequences. "Man may," he argued, "exercise a relative freedom in choosing how to use an invention, but if he uses it at all, it will entail its own social effect of which he is necessarily the servant" (1946: 10). All this implies a mechanical conception of cause and effect. Technology is the active agent, the "cause," and it interacts with people in a linear, often unidirectional, manner. The language used by this approach reflects this. The connection between technological innovation and what happens in society is compared variously to "the links in a chain" (Ogburn, 1957:22), or "the play on the billiard table" (Ogburn, 1957:20). A technical innovation is seen as having a direct or primary impact, which is usually the function it was intended to perform. Rippling out from this direct impact are a series of derivative, or secondary, effects which may in turn cause further derivative, or tertiary, effects, and so on (see Figure 7.1). Thus the causal relationship may become quite complex as one moves to more distant or higher order consequences. In other words, to speak of "social impact" is to speak in the language of technological determinism.

Figure 7.1

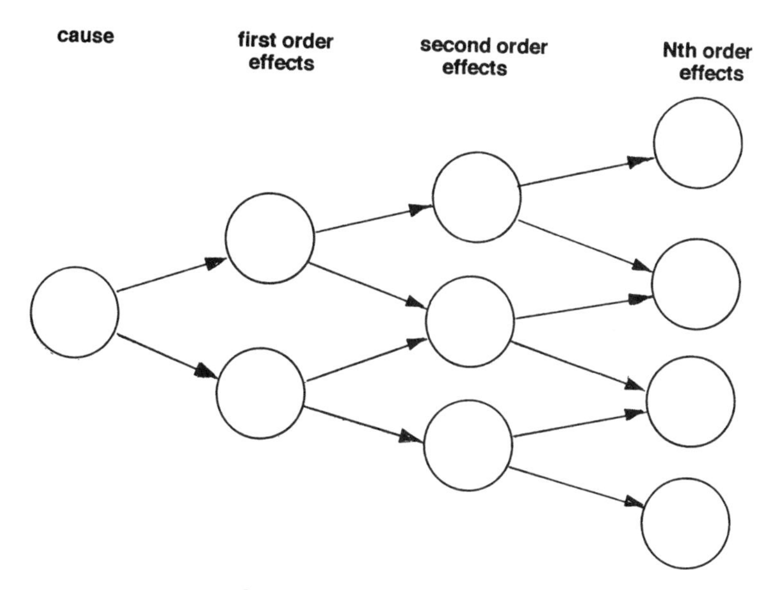

Ogburn's Social Impact Model

The second argument is that the pace of change is accelerating. Ogburn claimed that societies are "an accumulation of learned ways of behaving" (1964: 24). He put great emphasis on the state of the art as a determining factor in the pace of invention. An invention, he said, was the product of a combination of various existing elements. As material culture accumulates, there are more elements to be combined and the rate of invention increases. The state of the art of material culture thus acts in a way analogous to compound interest: "As the amount of interest paid an investor is a function of the size of the capital he has invested, so the number of inventions is a function of the size of the cultural base, that is, the number of existing elements in the culture" (Ogburn, 1964: 25). As a society develops, its material culture accumulates, the state of the art expands, the rate of technological innovation increases, and the velocity of social change accelerates. As a result, the pace of accumulation in the material culture accelerates even further, beginning the cycle again.

Third, Ogburn's school argued that since technology causes social change, and the pace of change is accelerating, a lag appears between technical innovation and cultural adjustment. As Ogburn put it: "Where one part of culture changes first, through some discovery or invention, and occasions changes in some part of culture dependent upon it, there frequently is a delay in the changes occasioned in the dependent part of the culture" (1950: 201). The reason for the lag was seen in the "obstacles" or "resistances" to technological innovation. These obstacles could be economic, psychological, or social (Nimkoff, 1957: 56-71). The consequences of cultural lag were seen as potentially disastrous. As Hornell Hart said, "If technology inevitably develops faster than social science, then the problems of social control will become ever more huge, while the means of control will progress far less rapidly than the need" (1957: 433-34).

This led to the fourth argument, that intervention is needed to hasten the adjustment to change. The Ogburn School pioneered what later would be called technology assessment. Perhaps the school's most significant work was Ogburn's *The Social Effects of Aviation* (1946), an examination of aviation not only in terms of transportation but in its effects on many social institutions, including the family, religion, education and health care. By anticipating change, they argued, obstacles and resistances can be overcome and society can adjust and adapt to technological change more quickly and smoothly.

Ogburn died in 1959 and his school soon ceased to exist, but it left a pervasive way of looking at technology and society. Alvin Toffler popularized their approach in books such as *Future Shock* (1970) and it was picked up by many advocates of technology assessment (TA) in the late 1960s. Obviously, Ogburn was not the only influence on the development of technology assessment. While many others contributed to the growing movement for TA, what they all shared was a belief in technological determinism. What changed in the sixties was a growing suspicion, doubt, and fear of technology as "technology is out of control" became a more common theme. Technology assessment gained impetus, then, not only for policy advisers concerned about long range planning but with many political groups as well, especially the growing environmental movement. As technology assessment became institutionalized in the 1970s it was divided into two components or subdisciplines, environmental impact assessment (EIA) and social impact assessment (SIA), the titles of which reflected the dominant understanding of the relationship between technology and change. EIA became the more institutionalized of the two, being required by law in some jurisdictions. SIA was never as well developed, although in the United States it too began to be institutionalized in the Congressional Office of Technology Assessment. The Reagan administration was hostile to technology assessment and in the 1980s the social movement for TA collapsed. So while Ogburn's work may be neglected today, his legacy continues in the way in which both critics and champions speak about technology and society.

Contradictions of Technology Assessment

Technology assessment was a procedure which never lived up to its promise. While there are a number of reasons for this, I contend that the key issue in the question of assessment is the question of values. It is in discourse on the relationship of technology and values that TA most displayed its weaknesses. Its problems were not so much technical or methodological but in establishing criteria, standards and evaluations. Only by understanding and interpreting this discourse and the assumptions which underlay it will we be able to find a replacement.

The question of the role of values has been fundamental to the assessment of technology. This is, of course, tautological, since assessment means evaluation. Values are part of the process by definition. Nevertheless, the role of values in assessment was always controversial. I think the problem lay in the assumptions the practitioners of TA made in doing their work.

A central assumption bequeathed to TA by Ogburn's school was a mechanistic conception of causality. The natural sciences were assumed to be the model for the social sciences. The "scientific method"—based upon induction; sharp dichotomies between data and theory, subject and object, and fact and value; and a strong emphasis upon the "objectivity" of results—was upheld as an ideal to be followed. Given their technological determinism, if one wanted to assess the "impact" of technology one had to use the most rigorous scientific methods to predict the secondary and tertiary effects which technology would have.

This assumption created two problems for TA. First was the attempt to separate "scientific fact" from "value-laden policy." Second was an inadequate conception of values. In both these cases overreliance upon a mechanical paradigm of science and technological determinism frustrated assessment.

The first problem, the attempt to separate "scientific fact" from "value-laden policy," arose from assumptions about validity. In their textbook on technology assessment, Porter et al. say validity "refers to being well grounded in fact and verifiable. In the case of research dealing with the future, we take validity to refer to the congruence between predicted and actual results" (1980: 45). Yet, they were forced to admit, predictions are always "uncertain." The inability to verify technology assessments arises in large part from trying to separate fact from value. That values are inherent and inescapable in the assessment process can be seen particularly clearly in two areas, the problem of setting the boundaries for the study and in the arguments over methodology.

First is the bounding or scoping problem, that is, the assumptions which set the limits and scope of the attempted assessment. Porter et al. noted that bounding "is difficult to accomplish, deeply intertwined with other assessment tasks, and crucial to the effective conduct and completion of an assignment" (1980: 66-67). They reported that in one analysis, half of the TAs studied were "severely impaired" (p. 67) by bounding problems. If boundaries are set too early or too narrowly, the TA may be irrelevant. If set too late or too broadly, it loses focus (not to mention what it does to the timetable and budget). Porter et al. listed six areas requiring bounding: time horizons; spatial extent; institutional involvements; technology and range of applications; impact sectors; and policy options (pp. 67-68). All six of these areas are premethodological, that is, they are not empirical questions to be answered but parameters of the study itself. In that they all require the practitioners to make assumptions and judgments, values were inherent in the process.

Additionally, the sponsors of a study usually have a say in setting boundaries. This makes the bounding problem a question of power. If the answers received depend upon the questions asked, who asks the questions? Who sets the agenda? Who defines terms? In other words, at its foundation assessment comes down to a question of *qui bono?* Whose good? At the most simple level this is seen in the acceptance or rejection of TAs by policy makers. As Berg and Rich reported: "One of the most significant factors found to affect utilization of TA studies was whether their potential users found them helpful or detrimental to the interests of their organizations" (in Boroush et al., 1980: 281). Such institutional self-interest can play a role from beginning to end in a study. Equally important are those questions which are left off the agenda. Unasked questions are just as revealing of assumptions and perspectives. Once more, values are inherent and inescapable in the process.

A second way in which values are inherent in the assessment process shows itself through the argument over methodology. Ever since Descartes, the mechanical paradigm of science has looked to methodology as a means of escaping bias and guaranteeing certainty. A wide variety of methods were developed for use in TA. From the beginning a heated debate raged between the advocates of quantitative and qualitative methodologies. Sherry Arnstein summed up these arguments:

> In addition to being unable to establish viable cause-and-effect relationships among the many variables involved, current quantification methods are not adequate for such unique tasks as (1) determining social acceptability and marketability for a new technology, (2) establishing, in advance, the dynamics of public attitudes toward future technological innovation, (3) anticipating changes in social values and perceptions 20-25 years hence, (4) tracing interactions among potential impacts of an innovation, and (5) distinguishing between facts and value based judgments. Having established that quantitative data alone are inadequate for these tasks, some of the practitioners were even more critical of current tools and methods for

> qualitative analysis, since these lack credibility, particularly to those scientist-experts who reject data that cannot be empirically validated.
> (in Boroush et al., 1980: 56-57)

These disputes were never resolved, in part because the claims made by the mechanical paradigm of science that their methodologies yield "objective" results with "predictive" power do not stand up to critical scrutiny. This is a problem which is structurally built into the TA process.

Prediction, or forecasting, was an essential element of technology assessment. The process "requires the projection of the future state of a technology so that its impact can be assessed" (Porter et al., 1980: 111). The record of social scientists as predictors, however, is incredibly bad. For instance, Furnas's (1936) notable study on the future of science totally missed, among other things, atomic bombs, computers, and space flight. Over the past twenty-five years, no one has been able to accurately predict the gyrations in the price of oil, in and of itself only one component of the economy. Needless to say, economic forecasts are most noted for being wrong. The record of futurologists is no better than that of science fiction writers. Most dramatically, of the thousands of "experts" in government, think tanks and universities, *no one* predicted the revolutions of 1989, the fall of communism, and the end of the cold war. We are simply unable to predict the future with any degree of accuracy, yet new, unforeseen, situations could have a tremendous effect on a technology being assessed.

The essence of the prediction problem is deeper, however. In spite of rigorous quantitative and qualitative methods, Porter et al. are forced to admit that "methodological sophistication contributes little to the accuracy of forecasts" (1980: 112). Simple extrapolations or judgment calls are as accurate as the most complex models and simulation. They add: "Basic core assumptions, not derivable from methodology, are the major determinants of forecast accuracy. When the core assumptions are valid, the choice of method is either obvious or secondary. When they are wrong, methodology can make little difference" (p. 112). They point out one additional factor. "The time horizon is the most consistent correlate of accuracy" (p. 113). In other words, it is easier to guess what is likely to happen next year than ten years hence. But this is a truism. The implication of what Porter et al. are saying[1] is that *values* rather than *method* is the key to "prediction." What counts are the paradigms the forecasters are working within. I think there are two reasons why this is so.

First is our poor record in prognosticating that which is totally new. The various methodologies for prediction rely upon extrapolations of present trends or simulation of the current system. As soon as something

[1] Which they apparently do not recognize themselves, since they go on to discuss methodology at great length.

new enters the picture, these sort of projections will be thrown off. Even relatively minor changes in the dynamics of a social system can dramatically distort forecasts. This problem is inherent in predictive methodology. Either the forecast is based solely on current data and trends, and therefore cannot encompass anything new, or assumptions about changes are introduced into the model by the researchers. But if that is done, the whole exercise is only as good as those assumptions. What counts is the adequacy of the forecaster's understanding of technological and social change and her or his imaginative abilities. One creative insight may have more validity than all the models based on the status quo.

Secondly, those making predictions forget that social science is a reflexive activity. As soon as recommendations are made they become part of the information upon which people base their decisions. Whatever inferences and extrapolations are made, whatever trends discerned, are liable to change simply because people now know about them. The TA itself becomes a part of the future which it was trying to predict. Inevitably, TA engages in a process of self-fulfilling or self-defeating prophecies.

In the end, the question of prediction is the wrong question. Prediction requires a determinate future, one in which technology is autonomous. Caught in Ogburn's view of the world, technology assessment could not escape its own internal contradiction: it was a branch of policy studies that in its most fundamental assumptions believed that machinery, not policy, is what really mattered. It could not see technology and society as a seamless web. If technology is a practice, the real problem shifts from foretelling to forthtelling, that is to say, to proclaiming the "signs of the times," the trends and tendencies, so that people can make purposeful and (reasonably) informed actions. The burden of prediction shifts from the arcanities of methodology to the values of the content. What matters is not accuracy of indicators but responsibility of recommendations. If we are to engage in an assessment process we must understand that we are not predicting the future but participating in creating it.

Through both the bounding problem and the argument over methodology, then, we can see that discussion of values is inherent in, and fundamental to, the assessment process. Without an adequate understanding of values, assessment lacks sufficient standards and criteria for evaluation. This is precisely the second problem of Ogburn's legacy to technology assessment: an inadequate conception of value.

Adherents of the mechanical paradigm of science assume that there is a clear distinction between facts and values and that when the researcher "begins to 'evaluate,' causal analysis almost always ceases—to the prejudice of the scientific results" (Max Weber, 1949: 33). Only empirical facts were seen as objective. When values were the object of study they were defined as "social facts," to be treated like any other empirical data.

The norm for the researcher was to remain "unbiased," "objective," and "non-judgmental." The emphasis of any study thus shifted to methodology, since only the most rigorous methods would be able to control for bias and ensure value-neutrality.

Since most of those who did technology assessment operated within the mechanical paradigm of science, these assumptions created problems for the assessment process. They were caught between their need for practical evaluation and their commitment to "scientific objectivity." Even those who recognized the importance and inescapability of value commitments to assessment were trapped in this dilemma. For example, Porter et al. went further than most in recognizing that "there is no precisely value-neutral position from which any assessment can be conducted" (1980: 45). But this insight was simply grafted onto their commitment to "follow established scientific principles and procedures" (p. 46). The most that they could do was call for "balance" and "recommend a *value-explicit* approach" (p. 45) in which "the assessors should try to spell out their assumptions and make clear their personal allegiances, so that the users of their study can judge the positions taken" (p. 45). Their understanding of values was fundamentally individualistic. Values were personal proclivities. They were seen as "choices" of the individual researcher which, like a hand of cards, are "held" and which can be "laid on the table." In their view this value-explicit approach was as necessary to overcome bias as was methodological rigour.

While this approach is far more sophisticated than the crude positivism still evident in much contemporary social science, it still is not satisfactory. The problem lies in their conception of values. One of the chief failings of technology assessment, the Organization for Economic Co-operation and Development (OECD) acknowledged, was that "relatively little attention has been paid to the fundamental point of technology assessment, i.e., a workable definition of the public interest in the name of which assessments are launched and carried out" (1983: 49). Lacking a conception of the common good, they were caught in a contradiction. Without a workable definition of the common good they had no standard of evaluation. But their notion of values as individual choices blocked such a definition. Recognizing this, they rather lamely responded: "Perhaps the most one can hope for, given that no neutral or generally valid definition of the common good seems possible ex-post, is that some kind of sufficient consensus will subsequently be achieved during the assessment process" (pp. 49-50). Such hopes can be most charitably described as naive.

The understanding of values held by the practitioners of TA had become the chief obstacle to overcoming the difficulties for which they introduced the discussion of values in the first place. Their language of values was inadequate. Echoing George Grant, Robert Bellah et al. argue:

> It should be clear by now that "values"...are in themselves no answer. "Values" turn out to be the incomprehensible, rationally indefensible thing that the individual chooses when he or she has thrown off the last vestige of external influence and reached pure, contentless freedom....The language of "values" as commonly used is self-contradictory precisely because it is not a language of value, or moral choice. It presumes the existence of an absolutely empty, unencumbered and improvisational self. It obscures personal reality, social reality, and particularly the moral reality that links person and society. (1985: 79-80)

An individualistic understanding of values cannot go beyond self-interest, the summing of private goods. But no calculus has ever been able to discover the common good by adding up private benefits. Only by actively searching for the good of the whole society can assessment transcend individual, organizational, communal, or class self-interest. As long as they remain committed to a paradigm which separates fact from value, subject from object, and thought from action, the practitioners of TA will never be able to achieve integration. As long as they try to make assessments from within a framework of assumptions which sees people and events interacting mechanically, their discussion of values will be unable to encompass the common good. As long as they define values as individualistic, even idiosyncratic, choices which may bias a study unless carefully explicated, values will never be more than a graft on their science.

Once we step outside the assumptions of the mechanical paradigm of science and of technological determinism, however, we open up the possibility of escaping this contradiction.[2] Then we can recognize that instead of individual specialists, each with a personal bag of "values," we are all members of communities engaged in social action. We are part of ongoing traditions. Critical examination of these traditions opens our discourse to the virtues of the community, to find what Ursula Franklin (1990) calls indivisible benefits. Such dialogue changes the definition of values and opens the possibility of a search for the common good.

From Assessment to Practice

Technology assessment failed because its practitioners believed that technology was autonomous, that its "impacts" determined the direction of social change, and because they did not have an adequate conception of

---

[2] By this I do *not* mean, however, that methodology is unimportant. Social science done without rigourous methodology is useless or even harmful. The crucial question is what kind of methods we use and the assumptions which underlie their use.

values. Evaluation of technology is crucial, but we have to go beyond assessment to a practice which embodies the six principles of redemptive technology.

When we look at the literature, there are many excellent sets of recommendations which show that there *are* alternatives. For instance (to mention just a few of the best) Pearl et al. (1990) give a comprehensive set of recommendations for removing barriers to women in computing which places emphasis on the necessity to address institutionalized practices even if they are not "technical." Lauren Wiener (1993) offers a list of seven questions to guide discussion about whether to adopt any digital system. Chandler et al. (1992) have an extensive list of recommendations for computerization of higher education. Menzies (1996) proposes a new social contract for the information highway. I have already mentioned some of Ursula Franklin's (1990) recommendations for a redemptive technology in chapter six. A major current undertaking is the ***Principles of Technorealism*** published on the World Wide Web. Its preamble states:

> Technorealism demands that we think critically about the role that tools and interfaces play in human evolution and everyday life. Integral to this perspective is our understanding that the current tide of technological transformation, while important and powerful, is actually a continuation of waves of change that have taken place throughout history....As technorealists, we seek to expand the fertile middle ground between techno-utopianism and neo-Luddism....We can be passionately optimistic about some technologies, skeptical and disdainful of others. Still, our goal is neither to champion nor dismiss technology, but rather to understand it and apply it in a manner more consistent with basic human values. (*Technorealism*, 1998)

Eight principles for dealing with issues surrounding the Internet follow, as well as an invitation to sign the manifesto and hot links to other sites discussing the issues. What all these recommendations have in common, unfortunately, is that few of them have been followed.

The failure to reform society, to practice justice and equality, to develop technology that benefits ordinary people and protects nature is not for lack of ideas. It is a question of who makes the decisions, and in whose interests are they made. In other words, it is the old question of *qui bono* again. Whose values do we incorporate, whose good is served? Liberals (and today's so-called conservatives) have a conception of values which only sees individual interests. They see the good as whatever the largest number of individuals want, while values are those goods which individuals choose. As we have already seen, this is an inadequate conception of value. While it is perhaps futile to abandon the language of values altogether (it is much too entrenched in our language) we need to give it substantive content. We need to build a new understanding of value as that which promotes the common good, which means we will have to shift the focus of ethical action from individuals to the community. Instead of only looking at the choices individuals make, we will have to look much more closely at what people do together. Our conception of the good will have to become embodied in institutions.

What *is* the common good? About 2700 years ago, a prophet answered by saying "Love justice, show mercy, and walk humbly with your God" (Micah 6:8). Both the Jewish sage Hillel and Jesus of Nazareth told their followers to "love your neighbour as yourself." The reciprocal relationships of life in community, with priority given to the needs of the poor and marginal, became a central principle of Western ethical thought. Now, there are few enough cases in empirical history of this principle actually being rigorously followed, but it always existed as a standard by which actions could be measured and judged. The history of the West is replete with prophets who arose to "speak truth to power." Only in the past two hundred years has this standard itself been eroded.[3]

Today, answers to "what is the common good?" start by taking a public opinion poll (if the question is asked at all). The good is no longer seen as an objective standard, but merely as whatever subjective opinions individuals hold at any given moment. The good as a criterion has been replaced by the market. Behaviour which may be appropriate in the economic sphere has, as Jürgen Habermas says (1981, 1987), colonized every aspect of our lives. The problem with this is that the market is neither democratic nor inclusive. In the marketplace, only those with wealth and power count. The poor and marginal have little influence, nature has no vote. The market is not an apparatus for meeting needs but only for satisfying greed. Because the market has no standards, even self-interest is circumscribed to that which is immediately profitable or gratifying. As a consequence, our communities are being undermined, the fabric of our society strained, and nature has become a "free good," there for the exploiting. "The final irony," says Robert Bellah et al., "is that this apparently economic conception of human life turns out to be profoundly destructive to our economy itself" (1991: 94). In the name of economic freedom we have diminished the quality of our lives and threatened the ecosystems upon which our prosperity, and even our existence, depend. When we substitute individual interests for the common good, Bellah

[3] The moral and ideological justification for a market-based industrial capitalism was laid in the eighteenth century by Bernard de Mandeville and Adam Smith. Both argued that the unregulated market, rather than any conception of the good, was the basis of the "wealth of nations." While today remembered as the founders of modern economics, both were first and foremost moral philosophers. See Dumont, 1977.

concludes, we ended up producing "a common bad that eventually erodes our personal satisfactions as well" (1991: 95).

So how do we find the common good and incorporate it into technology practice? One place to start is with the principles of redemptive technology. These are, of course, principles and not blueprints, guidelines rather than prescriptions for action. But as standards and criteria they can help to reform our practice of technology. For example, a Biblical standard of justice, in the context of modern information technology, would mean giving priority to users. The people who actually use the technology are all too often the forgotten people in system design. The virtuosity values of those who designed the technology will usually be represented, as will the economic values of those selling and implementing it. The needs of users, however, are rarely considered (Chandler et al., 1992; Pacey, 1983; Westrum, 1991). All too often, the aim of the technology is to replace people, deskill the work that remains, and allow management more centralized control (Menzies, 1996). The irony of IBM's old slogan, "Machines should work; people should think" is that across the industrialized world today, millions of workers are thinking about how to find a job. But high unemployment, and greater stress for those that remain, are not the inevitable by-products of information technology—they are the consequences of a particular technology practice which excluded user needs. Justice requires that user values be incorporated into technology practice. This does not mean that virtuosity or economic values need to be excluded, but rather that all three value sets must be incorporated and balanced. After all, it does not detract from the virtuosity of designers and engineers to develop systems which address users' needs. It might even be more of a challenge. Heather Menzies (1996: 152-54) provides an example of how computerization based on user values could have worked.

In the late 1980s the Vancouver Municipal and Regional Employees Union developed a counterplan to the city's design for restructuring. They engaged in a participatory research project which built upon municipal workers' experience and tacit knowledge while using technical experts as resources. The result was a plan for computerization which "would have both automated much of the routine administration (as planned) and extended their jobs even while extending the scope of service to the public" (p. 152). Their proposal included a registry of "at risk" people in special need of services, an inventory of trees on public land to guide public works in designing roads and pipelines, and retraining switchboard operators as information and referral agents linked to a database of all public services (p. 153). Unfortunately, management refused to negotiate and proceeded to use computerization to replace people. Menzies maintains that, although ultimately unsuccessful, the union's report shows what could be done if user values were taken seriously.

Justice may mean limiting the profits of corporations and the power of managers, but in the long run it could be rewarded by a healthier economy and a happier workforce. If we seek the good of all in our technology practice, rather than just the good of the wealthy and powerful, everyone could benefit. If we continue as we are, a few will benefit in the short run, but in the long term all will suffer.

Of course, we did not come to the present situation by accident. The powerful have created society as it is because they think (however short-sightedly) that it is in their interest to do so. We have already examined the process of appropriation by which social groups are able to incorporate their technological frames into the relations of ruling. With the stabilization and closure of debate over the meaning of a machine, the technology becomes a black box, surrounded by magic and mystification. If we are to create a redemptive technology practice we will have to reopen discourse about the meaning of technology. We have to take the lid off the black box. This involves four tasks.

First, we have to pay attention. Robert Bellah and his associates argue in *The Good Society* (1991) that much of modern life is distraction. Look, for example, at how much of today's discourse about computers is hyperbole and hype or how much of the Internet is occupied with spam, flame wars, idle chatter, and pointless web surfing. Look at how often we waste our time dreaming of utopias created by machinery. Look at how often we shrug off the consequences of technologial change with the expression "that's progress." We are distracted from what is important in life by amusements, enthusiasms, obsessions, and fears which fragment our time and leave us feeling stressed and depressed. In contrast to all the distraction thrust upon us, or the isolation of individual attempts to withdraw, Bellah says we have to pay attention, which "implies an openness to experience, a willingness to widen the lens of apperception when that is appropriate" (p. 257). We have to "concern ourselves with the larger meaning of things in the longer run rather than with short-term payoffs....A good society is one in which attention takes precedence over distraction" (p. 274). Paying attention is much more than a feverish effort to "stay informed" however (which itself can be just another form of distraction). "When we are giving our full attention to something," Bellah says, "when we are really attending, we are calling on all our resources of intelligence, feeling, and moral sensitivity" (p. 255). Spirituality is often at the heart of paying attention. Technological mysticism is distraction. Redemptive technology means paying attention. Once we start paying attention, mystification begins to dissolve and debates which have closed can begin to be reopened.

Reopening discussion about the meaning of technology will require, as a second task, that we dispel technological determinism. "Determinism," says Wiebe Bijker, "inhibits the development of democratic controls on technology because it suggests that all interventions are futile" (1995: 281). If the present use of technology is the only way it could be used, then those who benefit are both confirmed in their power and absolved of any responsibility. So a corporation which has used information technology to "downsize" its work force, for example, both collects its (short-term) profits and deflects blame from management to the machines. Critics are distracted and waste their efforts looking for solutions to the wrong problem. Reopened debate, using the tools of constructivism, ends the sense of inevitability around technological innovation and allows marginalized voices to be heard. Technology ceases to be the "all or nothing" choice it appears to be to those with low levels of inclusion (Bijker, 1995: 284)[4] and new possibilities are created. We can move from assessing what technology will do to us to practicing technology in a better future.

Technological determinism abdicates any sense of trust and responsibility. Our third task is to build both. In classical political philosophy, trust and responsibility were two virtues essential to citizenship. They still are. Robert Bellah and his associates observe: "Under modern conditions a society's economic and social development hinges essentially on ability to sustain institutions that mediate mutual trust and civic responsibility" (1991: 278). In North America today, however, cynicism and indifference are far more common than either responsibility or trust. This is hardly surprising in a society dominated by the market, where the watchwords are "what's in it for me?" and "let the buyer beware." This affects technology practice as much as it does elections. Just look at the licensing agreement on any package of software. Almost invariably, the manufacturer disclaims responsibility for how the product will perform. Buyers are just as cynical towards their obligations, turning software piracy into a multimillion dollar problem. Rebuilding trust and responsibility will not be easy, but without them redemptive technology is not possible. Life together in community depends upon trusting one another, and taking responsibility for our actions. But, in spite of the influence of the market and the corruption of elites, people still do trust each other, even if only in local and limited ways. But that is a beginning. These experiences can be fostered and encouraged and perhaps, just perhaps, they will grow. As Bellah concludes: "For none of us is it easy to override our mistrust and act

[4] Bijker argues that it is precisely low levels of inclusion that foster technological determinism in the first place.

responsibly in the universal community. Such a possibility is a gift; and when it comes, our response should be gratitude and celebration" (p. 286).

Fourth and finally, a redemptive technology practice will have to be embodied in institutions. People in North America have come to distrust their institutions, particularly in the United States and increasingly so in Canada. The very word "institution" evokes images of callous executives, corrupt politicians, and mindless bureaucracy. These images are far different from a sociological understanding of institutions, which are defined as "a pattern of expected action of individuals or groups enforced by social sanctions, both positive and negative" (Bellah et al., 1991: 10). Institutions are far more than organizations. They are the normative patterns which constitute our communities, and as such are fundamentally moral. As Bellah and his associates summarize:

> In our life with other people we are engaged continuously, through words and actions, in creating and re-creating the institutions that make life possible. This process is never neutral but is always ethical and political, since institutions (even such intimate an institution as the family) live or die by ideas of right and wrong and conceptions of the good. Conversely, while we in concert with others create institutions, they also create us: they educate us and form us—especially through the socially enacted metaphors they give us, metaphors that provide normative interpretations of situations and actions. The metaphors may be appropriate or inappropriate, but they are inescapable. (p. 11)

Institutions are reflexive. As the patterns of meaning which organize our relationships with each other, they grow out of our discourse while shaping our practices, which in turn shapes our discourse (cf. Berger and Luckmann, 1966; Giddens, 1984). Life in community only exists in and through our institutions.

What matters, then, is the kind of institutions in which we want to live. As the embodiment of our practices, we can have institutions which offer us distractions, or ones which help us to pay attention. We can have institutions which abdicate responsibility and betray trust, or ones which foster both. We can have institutions which proclaim that technology is only about power and control, or ones which emphasize creativity and respect for limits. Our institutions can drain values of content, or nurture the common good. They can embody technological mysticism, or they can build a redemptive technology.

An Invitation to Talk

I began this chapter with the perennial question, what is to be done? Those who were expecting a step-by-step plan are by now disappointed. Redemptive technology is a practice, not a utopia. Unlike some, I cannot make confident predictions based on faith in the power and inevitability of technological change. I cannot offer reassurances that experts somewhere have the answers and are ready to implement them. I cannot even offer a definitive technique for assessing technology. What I do offer is an invitation to talk.

If we are to be wise in our development and use of technology we must assess and evaluate it, but what is necessary for assessment is not primarily methodology but discourse. We have to talk about what we want and need, and how (or whether) technology may aid us in achieving it. But most of all we have to talk about what is good and about what would be just. We have to be creative in our discussion, and we also have to discover and respect the limits to what technology can do. We have to bring others into our discussions, and seek to participate in theirs, so that our understanding of the good reciprocates theirs. We have to aim for a technology practice that includes everyone, uniting our local communities while addressing global concerns.

Even this may sound utopian. When would we find the time? Part of the answer is to pay attention and avoid being distracted. A bigger part, however, is to be more thoughtful as we engage in the discussions of technology which are already going on around us. We can begin wherever we are—in our workplaces, schools, universities, churches, or union halls. One obvious place is the Internet. In spite of premature claims by its enthusiasts, whether or not the Internet can grow into a redemptive technology remains to be seen. There is a qualitative difference between real dialogue and the fragmentation, chatter, and abusive flame wars so characteristic of the net today. Anarchy and democracy are not the same thing. On the other hand, a number of moderated discussion groups and Web sites offer real promise of meaningful dialogue. *Feed* magazine's online discussion of the *Principles of Technorealism* (*Feed*, 1998) is an outstanding example.

We should expect disagreements, even major conflicts. People will come to the debate with different backgrounds and assumptions and even if they put aside "what's good for me," there will be real and legitimate differences of interpretation and application of principles. But this will always be the case. Debate is not a conflict-free utopia. What we have to do is trust each other, take responsibility for our actions, and agree to keep talking and working together.

Today technological mysticism is the One True Faith, the dominant way of talking about technology. It identifies knowledge with power and technology with control. It has distorted our understanding, perpetuated harmful myths, and encouraged a destructive use of technology. The alternative is a redemptive technology practice that sees a seamless web of technical artifacts, organizations, and cultural understandings, and which aims for a good society. Instead of seductive promises of a technical fix, it offers discussion and hard work. But from discussion can come action, and action can build tomorrow.

REFERENCES

Adelson, Rachel K. 1997. "Computer Comedy ToNite." *Livewire: Newsletter of the Association for Women in Computing*. www.awc-hq.org/livewire/199703.html.

Adelson, Rachel K. 1996. "Computer Toys R Us Vs Them." *Livewire: Newsletter of the Association for Women in Computing*. www.awc-hq.org/livewire/199612.html.

Adorno, Theodor. 1974. "The Stars Down to Earth: The Los Angeles Times Astrology Column." *Telos* 19: 13-90.

Advocates for Women in Science, Engineering and Mathematics [AWSEM]. 1997. www.awsem.com/gender.html.

Alcoff, Linda and Elizabeth Potter (eds.). 1993. *Feminist Epistemologies.* New York: Routledge.

Alves, Rubem. 1972. *Tomorrow's Child: Imagination, Creativity, and the Rebirth of Culture.* New York: Harper & Row.

Anderson, R. 1987. "Females Surpass Males in Computer Problem Solving: Findings from the Minnesota Computer Literacy Assessment." *Journal of Educational Computing Research* 3(1): 39-51.

Bacon, Francis. 1620/1960. *The New Organon and Related Writings,* edited by Fulton Anderson. New York: The Liberal Arts Press.

Bailey, Edward. 1990. "The Implicit Religion of Contemporaty Society: Some Studies and Reflections." *Social Compass* 37(4): 483-97.

Bailey, Edward. 1983. "The Implicit Religion of Contemporary Society: An Orientation and Plea for its Study." *Religion* 13: 69-83.

Barbour, Ian. 1993. *Ethics in an Age of Technology.* San Francisco: Harper.

Barbour, Ian. 1990. *Religion in an Age of Science.* San Francisco: Harper.

Barron, Daniel. 1996. "It Began with Hypatia and Ada; Helping Girls Bridge the Technology Gender Gap." *School Library Media Activities Monthly* 12(7): 47-50.

Becker, H. J., and C. W. Sterling. 1987. "Equity in School Computer Use: National Data and Neglected Considerations." *Journal of Educational Computing Research* 3(3): 289-311.

Belknap, P. and W. Leonard. 1991. "A Conceptual Replication and Extension of Erving Goffman's Study of Gender Advertisements." *Sex Roles* 25 (3/4): 103-18.

Bell, Daniel. 1973. *The Coming of Post-Industrial Society.* New York: Basic Books.

Bellah, Robert (ed.). 1973. *Emile Durkheim on Morality and Society.* Chicago: University of Chicago Press.

Bellah, Robert. 1970. *Beyond Belief: Essays on Religion in a Post-Traditional World.* New York: Harper & Row.

Bellah, Robert, Richard Masden, William Sullivan, Ann Swidler, and Steven Tipton. 1991. *The Good Society.* New York: Alfred A. Knopf.

Bellah, Robert; Richard Masden, William Sullivan, Ann Swidler, and Steven Tipton. 1985. *Habits of the Heart: Individualism and Commitment in American Life.* Berkeley: University of California Press.

Berg, Anne-Jorunn and Merete Lie. 1995. "Feminism and Constructivism: Do Artifacts Have Gender?" *Science, Technology, and Human Values* 20(3): 332-51.

Berger, Peter and Thomas Luckmann. 1966. *The Social Construction of Reality.* Garden City, N.Y.: Doubleday.

Berghahn, Klaus L. 1987. "George Johann Heinrich Faust: The Myth and Its History." In *Our Faust? Roots and Ramifications of a Modern German Myth,* edited by Reinhold Grimm and Jost Hermand. Madison: The University of Wisconsin Press.

Bernstein, Richard J. 1983. *Beyond Objectivism and Relativism.* Philadelphia: University of Pennsylvania Press.

Bijker, Wiebe. 1995. *Of Bicycles, Bakelites, and Bulbs: Towards a Theory of Sociotechnical Change.* Cambridge, Mass.: MIT Press.

Bijker, Wiebe. 1993. "Do Not Despair: There is Life After Constructivism." *Science, Technology, and Human Values* 18 (1): 113-38.

Bijker, Wiebe. 1987. "The Social Construction of Bakelite: Toward a Theory of Invention." In *The Social Construction of Technological Systems,* edited by W. Bijker, T. Hughes, and T. Pinch. Cambridge, Mass.: MIT Press

Bijker, Wiebe and John Law (eds.). 1992. *Shaping Technology/Building Society.* Cambridge, Mass.: MIT Press.

Bijker, W., T. Hughes, and T. Pinch (eds.). 1987. *The Social Construction of Technological Systems.* Cambridge, Mass.: MIT Press

Bijker, W. and T. Pinch. 1987. "The Social Construction of Facts and Artifacts: Or How the Sociology of Science and the Sociology of Technology Might Benefit Each Other." In *The Social Construction of Technological Systems,* edited by W. Bijker, T. Hughes, and T. Pinch. Cambridge, Mass.: MIT Press

Biraimah, Karen. 1993. "The Non-Neutrality of Educational Computer Software." *Computers and Education* 20(4): 283-90.

Bloch, Ernst. 1988. *The Utopian Function of Art and Literature,* trans. by Jack Zipes and Frank Mecklenburg. Cambridge, Mass.: MIT Press.

Bloch, Ernst. 1959/1986. *The Principle of Hope,* trans. by Neville Plaice, Stephan Plaice, and Paul Knight. Cambridge, Mass.: MIT Press.

Bloor, David. 1976. *Knowledge and Social Imagery.* London: Routledge & Kegan Paul.

Bolter, J. David. 1984. *Turing's Man: Western Culture in the Computer Age.* Chapel Hill: University of North Carolina Press.

Boot, William. 1986. "NASA and the Spellbound Press." *Columbia Journalism Review* (July/August): 23-29.

Bouroush, Mark, Ken Chen, and Alexander Christakis (eds.). 1980. *Technology Assessment: Creative Futures.* New York: North Holland.

Brecher, D. 1989. "Gender and Learning: Do Women Learn Differently?" In *Women, Work and Computerization: Forming New Alliances,* edited by K. Tijdens, M. Jennings, I.Wagner, and M. Weggelaar. Amsterdam: North-Holland.

Bretl, D. and J. Cantor. 1988. "The Portrayal of Men and Women in U.S. Television Commercials: A Recent Content Analysis and Trends Over 15 Years." *Sex Roles* 18 (9/10): 595-609.

Brombacher, Bruce E. 1982. "Forward." In *Computers in the Classroom,* edited by Henry S. Kepner. Washington, D.C.: National Education Association.

Brough, Neil. 1994. *New Perspectives of Faust: Studies in the Origins and Philosophy of the Faust Theme in the Dramas of Marlowe and Goethe.* Frankfurt am Main: Peter Lang.

Brown, Gillian and George Yule. 1983. *Discourse Analysis.* Cambridge: Cambridge University Press.

Callon, M. 1987. "Society in the Making: The Study of Technology as a Tool for Sociological Analysis." In *The Social Construction of Technological Systems,* edited by W. Bijker, T. Hughes, and T. Pinch. Cambridge, Mass.: MIT Press

Chandler, David, Barbara Kaltz, Murray Knuttila, R. Brien Maguire, William Stahl, and Larry Symes. 1992. *Computer Technology for Higher Education Vol. III. The Canadian Experience.* New Delhi: Concept Publishing.

Chappell, Kelly. 1996. "Mathematics Computer Software Characteristics with Possible Gender-Specific Impact: A Content Analysis." *Journal of Educational Computing Research* 15(1): 25-35.

Charp, Sylvia, W. Bozeman, H. Altschuler, R. D'Orazio, and D. Spuck. 1982. *Layman's Guide to the Use of Computers in Education.* Washington, D.C.: The Association for Educational Data Systems.

Chen, M. 1987. "Gender Differences in Adolescent's Uses of and Attitudes Toward Computers." In *Communications Yearbook 10,* edited by M. McLauglin. Beverly Hills: Sage Publishing.

Churchland, Paul M. and Patricia Smith Churchland. 1990. "Could A Machine Think?" *Scientific American* 262(1): 32-37.

Clarke, Arthur C. 1973. *Profiles of the Future.* New York: Harper & Row.

Clarke, V. 1986. "Why Are Girls Under-Represented? Suggestions from the Literature." *Australian Educational Computing* 1:46-50.

Clarke, V. and S. Chambers. 1989. "Gender-Based Factors in Computing Enrollments and Achievement: Evidence from a Study of Tertiary Students." *Journal of Educational Computing* 5 (4): 409-29.

Cohen, Norman. 1970. *The Pursuit of the Millenium.* New York: Oxford University Press.

Collins, Harry. 1985. *Changing Order: Replication and Induction in Scientific Practice.* London: Sage Publications.

Collins, Harry and Trevor Pinch. 1993. *The Golem: What Everyone Should Know About Science.* Cambridge: Cambridge University Press.

Computer Professionals for Social Responsibility. 1995. www.cpsr.org/program/gender/index.html.

Corn, Joseph (ed.). 1986. *Imagining Tommorrow: History, Technology, and the American Future.* Cambridge, Mass.: MIT Press.

Corn, Joseph J. and Brian Horrigan. 1984. *Yesterday's Tomorrows: Past Visions of the American Future.* New York: Summit Books.

Crenshaw, James L. 1981. *Old Testament Wisdom: An Introduction.* Atlanta: John Knox Press.

Culley, L. 1990. "Option Choice and Careers Guidance: Gender and Computing in Secondary Schools." *Western European Education* 21(4): 41-53.

Dambrot, F., M. Watkins-Malek, S. M. Silling, R. Marshall, and J. A. Garver. 1985. "Correlates of Sex Difference in Attitudes Toward and Involvement with Computers." *Journal of Vocational Behavior* 27(1): 71-86.

Danzinger, James N. 1985. "Social Science and the Social Impacts of Computer Technology." *Social Science Quarterly* 66 (1).

de Bolt, Joseph. 1969. "Belief Systems and Evolution: A Distinction Between Magic and Religion and Its Implication for Sociocultural Change." *Canadian Review of Sociology and Anthropology* 6(2): 80-91.

Demetrulias, D. and N. Rosenthal. 1985. "Discrimination Against Females and Minorities in Microcomputer Advertising." *Computers and the Social Sciences* 1: 91-95.

Deringer, Dorothy K. and Andrew R. Molnar. 1982. "Key Components for a National Computer Literacy Program." In *Computer Literacy,* edited by Robert Seidel, Ronald Anderson, and Beverly Hunter. Orlando, Fl.: Academic Press.

Dewdney, A. K. 1988. "Nanotechnology: Wherein Molecular Computers Control Tiny Circulatory Submarines." Computer Recreations. *Scientific American* 258(1): 100-103.

Dickson, Gordon R. 1984. *The Final Encyclopedia.* New York: Berkeley Publishing.

Doniol-Shaw, G. 1989. "Technical Culture, Training and Female Employment." In *Women, Work and Computerization: Forming New Alliances,* edited by K. Tijdens, M. Jennings, I.Wagner, and M. Weggelaar. Amsterdam: North-Holland.

Downs, A. and S. Harrison. 1985. "Embarrassing Age Spots or Just Plain Ugly? Physical Attractiveness Stereotyping as an Instrument of Sexism on American Television Commercials." *Sex Roles* 13 (1/2): 9-19.

Drexler, K. Eric. 1986. *Engines of Creation.* Garden City, N.Y.: Doubleday

Drucker, Peter. 1968. *The Age of Discontinuity.* New York: Harper & Row.

Dublin, Max. 1989. *Futurehype: The Tyranny of Prophecy.* Markham, Ont.: Viking.

Dumont, Louis. 1977. *From Mandeville to Marx: The Genesis and Triumph of Economic Ideology.* Chicago: University of Chicago Press.

Durkheim, Emile. 1924/1953. *Sociology and Philosophy,* trans. by D. Pocock. Glencoe, Il.: The Free Press.

Durkheim, Emile. 1915. *The Elementary Forms of the Religious Life,* trans. by J. W. Smith. New York: Macmillan.

Edge, David. 1995. "Reinventing the Wheel." In *Handbook of Science and Technology Studies,* edited by S. Jasanoff, G. Markle, J. Petersen, and T. Pinch. Thousand Oaks: Sage Publications.

Eisenberg, Anne. 1997. "Disliking the Internet." *Scientific American.* 276(6): 44.

Eliade, Mircea. 1978. *The Forge and the Crucible,* trans. by Stephen Corrin. Chicago: University of Chicago Press.

Eliade, Mircea. 1963. *Myth and Reality,* trans. by Willard Trask. New York: Harper & Row.

Elkjær, B. 1989. "Myth and Reality About Women and Technology." In *Women, Work and Computerization: Forming New Alliances,* edited by K. Tijdens, M. Jennings, I.Wagner, and M. Weggelaar. Amsterdam: North-Holland.

Elkjær, B. 1986. "The New Technology is Women's Opportunity." In *Women Challenge Technology.* European Conference on Women, Natural Sciences and Technology, edited by M. Dahms et al. Elsinore, Denmark. University of Aalborg: Centertrykkeriet.

Ellul, Jacques. 1980. *The Technological System,* trans. by Joachim Neugroschel. New York: Continuum.

Ellul, Jacques. 1964. *The Technological Society* (rev. ed.), trans. by John Wilkinson. New York: Random House.

Etzkowitz, H., C. Kemelgor, M. Neuschatz, and B. Uzzi. 1994. "Barriers to Women in Academic Science and Engineering." In *Who Will Do Science? Educating the Next Generation,* edited by W. Pearson and I. Fletcher. Baltimore: Johns Hopkins University Press.

Feed. 1998. www.feedmag.com/html/dialog/98.03dialog/98.03dialog_ master.html.

Feigenbaum, Edward and Pamela McCorduck. 1983. *The Fifth Generation.* Reading, Mass.: Addison-Wesley Publishing.

Ferré, Frederick. 1993. *Hellfire and Lightning Rods: Liberating Science, Technology, and Religion.* Maryknoll: Orbis Books.

Fischer, William, Willard Hamilton, Curtis McLaughlin and Robert Zmud. 1986. "The Elusive Product Champion." *Research Management* (May-June): 13-16.

Fleck, Ludwik. 1935/1979. *Genesis and Development of a Scientific Fact,* trans. by T. Trenn. Chicago: University of Chicago Press.

Fowler, C. J. H. and D. Murray. 1987. "Gender and Cognitive Style Differences at the Human-Computer Interface." In *Human-Computer Interaction—INTERACT '87,* edited by Bullinger and Shackel. Amsterdam: North-Holland.

Franklin, Ursula. 1990. *The Real World of Techhnology.* Montreal: CBC Enterprises.

Frenkel, K. 1990. "Women and Computing." *Communications of the ACM 33*(11): 36-46.

Funk, Jeanne and Debra Buchman. 1996. "Playing Violent Video and Computer Games and Adolescent Self-Concept." *Journal of Communication* 46(2): 19-32.

Furnas, C. C. 1936. *The Next Hundred Years: The Unfinished Business of Science.* New York: Reynal and Hitchcock.

Geertz, Clifford. 1973. *The Interpretation of Cultures.* New York: Basic Books.

Gerhardt, Mary and Allan Russell. 1984. *Metaphoric Process: The Creation of Scientific and Religious Understanding.* Fort Worth: Texas Christian University Press.

Gibbs, W. Wayt. 1997. "Taking Computers to Task." *Scientific American* 227(1): 82-89.

Giddens, Anthony. 1984. *The Constitution of Society.* Berkeley: University of California Press.

Gill, Rosalind. 1996. "Power, Social Transformation, and the New Determinism: A Comment on Grint and Woolgar." *Science Technology and Human Values* 21(3): 347-53.

Gitlin, Todd. 1980. *The Whole World is Watching.* Berkeley: University of California Press.

Goethe, Johann Wolfgang. 1962. *Faust Vol. I The Prologues and Part One, Vol. II Part Two,* the Bayard Taylor translation, revised and edited by Stuart Atkins. New York: Collier Books.

Gould, S. 1987. "Gender Differences in Advertising Response and Self-Consciousness Variables." *Sex Roles* 16 (5/6): 215-25.

Grant, George. 1986. *Technology and Justice.* Toronto: House of Anansi.

Grant, George. 1969. *Technology and Empire.* Toronto: House of Anansi.

Grant, George. 1965. *Lament For A Nation.* Toronto: McClelland & Stewart.

Greenbaum, J. 1987. *The Head and the Heart.* Aarhus University, Denmark: Department for Information and Media Science.

Gregory, Bruce. 1990. *Inventing Reality: Physics as Language.* New York: John Wiley and Sons.

Griffiths, M. 1988. "Strong Feelings About Computers." *Women's Studies International Forum.* 11 (2): 145-54.

Grimm, the Brothers. 1945. *Grimm's Fairy Tales,* trans. by E.V. Lucas, L. Crane and M. Edwards. New York: Gosset and Dunlop.

Grint, Keith and Steve Woolgar. 1996. "A Further Decisive Refutation of the Assumption That Political Action Depends on the 'Truth' and a Suggestion That We Need to Go Beyond This Level of Debate: A Reply to Rosalind Gill." *Science, Technology and Human Values* 21(3): 354-57.

Grint, Keith and Steve Woolgar. 1995. "On Some Failures of Nerve in Constructivist and Feminist Analyses of Technology." *Science, Technology and Human Values* 20(3): 286-310.

Grossman, Wendy. 1998. "Access Denied." *Scientific American* 279 (2) (August): 38.

Gunn, Cathy. 1994. "Development of Gender Roles: Technology as an Equity Strategy." In *Recreating the Revolution: Proceedings of the Annual National Educational Computing Conference.* Boston, MA. Syracuse, N.Y.: ERDS.

Habermas, Jürgen. 1991. *Moral Consciousness and Communicative Action,* trans. by Christian Lenhardt and Shierry Nicholsen. Cambridge, Mass.: MIT Press.

Habermas, Jürgen. 1987. *The Theory of Comunicative Action Volume Two Lifeworld and System: A Critique of Functionalist Reason,* trans. by Thomas McCarthy. Boston: Beacon Press.

Habermas, Jürgen. 1981. *The Theory of Comunicative Action Volume One Reason and the Rationalization of Society,* trans. by Thomas McCarthy. Boston: Beacon Press.

Haile, H. G. 1965. *The History of Doctor Johann Faustus.* Urbana, Ill.: University of Illinois Press.

Halfhill, Thomas R. 1997. "Cheaper Computing, Part I." *Byte* 22(4): 66-80.

Hall, Edward T. 1969. *The Hidden Dimension.* Garden City: Doubleday.

Hamilton, Edith. 1942. *Mythology.* Boston: Little, Brown and Co.

Hanegraaff, Wouter. 1997. *New Age Religion and Western Culture: Esotericism in the Mirror of Secular Thought.* Albany: State University of New York Press.

Haraway, Donna. 1991. *Simian, Cyborgs. and Women: The Reinvention of Nature.* New York: Routledge.

Harding, Sandra. 1991. *Whose Science? Whose Knowledge? Thinking from Women's Lives.* Ithaca, N.Y.: Cornell University Press.

Harding, Sandra. 1986. *The Science Question in Feminism.* Ithaca: Cornell University Press.

Harding, Sandra. 1976. *Can Theories Be Refuted? Essays on the Duhem-Quine Thesis.* Boston: Dordrecht-Holland.

Hart, Hornell. 1957. "The Hypotheses of Cultural Lag." In *Technology and Social Change,* edited by Francis Allen et al. New York: Appleton-Century-Crofts.

Hawkins, J. 1985. "Computers and Girls: Rethinking the Issues." *Sex Roles* 13(3/4): 165-80.

Heelas, Paul. 1996. *The New Age Movement: The Celebration of the Self and the Sacralization of Modernity.* Oxford: Blackwell.

Hodes, Carol. 1996. "Gender Representations in Mathematics Software." *Journal of Educational Technology Systems* 24(1): 67-73.

Holmes, David (ed.). 1997. *Virtual Politics: Identity and Community in Cyberspace.* Thousand Oaks, Calif.: Sage Publications.

Horkheimer, Max. 1972. *Critical Theory,* trans. by M. J. O'Connell and others. New York: Seabury.

Horkheimer, Max. 1947/1974. *Eclipse of Reason.* New York: Seabury.

Horkhiemer, Max and Theodor W. Adorno. 1944/1972. *Dialectic of Enlightenment,* trans. by John Cumming. New York: Seabury.

Høyrup, E. 1986. "Science and Feelings." In *Women Challenge Technology. European Conference on Women, Natural Sciences and Technology,* edited by M. Dahms et al. Elsinore, Denmark. University of Aalborg: Centertrykkeriet.

Hunter, Beverly. 1982. "Computer Literacy: 1949-1979." In *Computer Literacy,* edited by Robert Seidel, Ronald Anderson, and Beverly Hunter. Orlando, Fl.: Academic Press.

Inose, Hiroshi and John R. Pierce. 1984. *Information Technology and Civilization.* New York: W.H. Freeman.

Jacobs, J. 1988. "Social Implications of Computers: Ethical and Equity Issues." *ACM SIGCUE Outlook* 20 (1): 100-14.

Jennings, M. 1986. "Teaching Computing to Women—A Feminist Approach." In *Women Challenge Technology. European Conference on Women, Natural Sciences and Technology,* edited by M. Dahms et al. Elsinore, Denmark. University of Aalborg: Centertrykkeriet.

Keller, Evelyn Fox. 1992. *Secrets of Life, Secrets of Death.* New York: Routledge.

Keller, Evelyn Fox. 1985. *Reflections on Gender and Science.* New Haven and London: Yale University Press.

Kidder, Tracy. 1981. *The Soul of a New Machine.* New York: Avon Books.

Kiesler, S., L. Sproull, and J. Eccles. 1985. "Pool Halls, Chips and War Games: Women in the Culture of Computing." *Psychology of Women Today* 9(Dec): 451-62.

Klassen, Daniel L. and Ronald E. Anderson. 1982. "Computer Literacy." In *Computers in the Classroom,* edited by Henry S. Kepner, Washington, D.C.: National Education Association.

Knorr-Cetina, Karin. 1981. *The Manufacture of Knowledge.* Oxford: Pergamon Press.

Kolakowski, Leszek. 1968. *Toward A Marxist Humanism,* trans. by Jane Peel. New York: Grove Press.

Kramarae, Cheris (ed.). 1988. *Technology and Women's Voices.* New York: Routledge & Kegan Paul.

Kramer, Kevin and Nancy Knupfer. 1997. "Gender Equity in Advertising on the World Wide Web: Can It Be Found?" In *Proceedings of Selected Research and Development Presentations at the 1997 National Convention of the Association for Educational Communications and Technology.* Albuquerque, N.M. Syracuse, N.Y.: EDRS.

Knupfer, Nancy. 1997. "New Technologies and Gender Equity: New Bottles With Old Wine." In *Proceedings of Selected Research and Development Presentations at the 1997 National Convention of the Association for Educational Communications and Technology.* Albuquerque, NM. Syracuse, N.Y.: EDRS.

Kuhn, Thomas. 1970 . *The Structure of Scientific Revolutions, second edition.* Chicago: University of Chicago Press.

Lakoff, George and Mark Johnson. 1980. *Metaphors We Live By.* Chicago: University of Chicago Press.

Landauer, Thomas K. 1995. *The Trouble With Computers.* Cambridge, Mass.: MIT Press.

Latour, Bruno. 1996. *Aramis or the Love of Technology,* trans. by Catherine Porter. Cambridge, Mass.: Harvard University Press.

Latour, Bruno. 1993. *We Have Never Been Modern,* trans. by Catherine Porter. Cambridge, Mass.: Harvard University Press.

Latour, Bruno. 1987. *Science in Action.* Cambridge, Mass.: Harvard University Press.

Latour, Bruno and Steve Woolgar. 1979. *Laboratory Life: The Construction of Scientific Facts.* Princeton: Princeton University Press.

Lee, Kar-tin. 1995. "Teachers' Computer Use Patterns: Factors Which Influence the Degree of Integration of Computing into Their Professional and Private Lives." *Gates* 2(1): 8-17.

Lee, Martin and Norman Solomon. 1991. *Unreliable Sources: A Guide to Detecting Bias in News Media.* New York: Carol Publishing Group.

Leiss, William. 1990. *Under Technology's Thumb.* Montreal: McGill-Queen's University Press.

Levin, Harry. 1966. "Science Without Conscience." In *Christopher Marlowe's Doctor Faustus: Text and Major Criticism*, edited by Irving Ribner. New York: Odyssey Press.

Lewis, Harold W. 1980. "The Safety of Fission Reactors." *Scientific American* 242 (3): 53-65.

Lewis, L. 1985. "New Technologies, Old Patterns: Changing the Paradigm." *Educational Horizons* 63(3): 129-32.

Lindelow, John. 1983. *Administrators Guide to Computers in the Classroom.* Eugene, Oregon: Clearinghouse for Educational Management, University of Oregon.

Linn, M. 1985. "Fostering Equitable Consequences from Computer Learning Environments." *Sex Roles* 13(3/4): 229-40.

Littleton, Karen. 1993. "Gender and Software Effects in Computer-Based Problem Solving." Paper Presented at the Annual Meeting of the Society for Research in Child Development New Orleans, LA. Syracuse, N.Y.: ERDS.

Logan, Robert. 1977. *Nothing But the Facts: A Critical Analysis of Science Reporting in American Newspapers*. Unpublished Ph.D. dissertation. University of Iowa.

Lowrance, William W. 1985. *Modern Science and Human Values*. New York: Oxford University Press.

Luckmann, Thomas. 1967. *The Invisible Religion.* New York: Macmillan.

MacDougall, John. 1913/1973. *Rural Life in Canada.* Toronto: University of Toronto Press.

MacKenzie, Donald and Judy Wajcman (eds.). 1985. *The Social Shaping of Technology.* Milton Keynes, England: Open University Press.

Maclean's. 1979-1988. Vol. 82 to Vol. 101.

Mahoney, Judy and Nancy Knupfer. 1997. "Language, Gender and Cyberspace: Pulling the Old Stereotypes into New Territory." In *Proceedings of Selected Research and Development Presentations at the 1997 National Convention of the Association for Educational Communications and Technology.* Albuquerque, NM. Syracuse, N.Y.: EDRS.

Makrakis, Vasilios and Toshio Sawada. 1996. "Gender, Computers and Other School Subjects Among Japanese and Swedish Students." *Computers and Education* 26(4): 225-31.

Malinowski, Bronislaw. 1925/1948. *Magic, Science and Religion,* intro. by R. Redfield. Garden City, NY: Doubleday.

Mangan, J. Marshall. 1992. "The Politics of Computer Literacy in Ontario Schools." Paper presented at the annual meetings of the Canadian Sociological and Anthropological Association, Charlottetown, Prince Edward Island, Canada.

Mangione, Melissa. 1995. "Understanding the Critics of Educational Technology: Gender Inequities and Computers, 1983-1993." In *Proceedings of the 1995 Annual National Convention of the Association for Educational Communications and Technology.* Anaheim, CA. Syracuse, N.Y.: EDRS.

Mannheim, Karl. 1936. *Ideology and Utopia,* trans. by Louis Wirth and Edward Shils. New York: Harcourt Brace and World.

Manuel, Frank (ed.). 1966. *Utopias and Utopian Thought.* Boston: Houghton Mifflin.

Manuel, Frank and Fritzie Manuel. 1979. *Utopian Thought in the Western World.* Cambridge, Mass.: Harvard University Press.

Marcuse, Herbert 1964. *One Dimensional Man*. Boston: Beacon Press.

Marlowe, Christopher. 1966. "The Tragicall History of the Life and Death of Doctor Faustus." In *Christopher Marlowe's Doctor Faustus: Text and Major Criticism,* edited by Irving Ribner. New York: Odyssey Press.

Marshall, J. P., F. Erickson, and J. Vonk. 1986. "Access Equity of Computer Instruction in High Schools." *Free Inquiry in Creative Sociology* 14(2): 129-31.

Marvin, Carolyn. 1986. "Dazzling the Multitude: Imagining the Electric Light as a Communications Medium." In *Imagining Tommorrow: History, Technology, and the American Future*, edited by Joseph Corn. Cambridge, Mass.: MIT Press.

Masuda, Yoneji. 1980. *The Information Society*. Washington, D.C.: World Future Society.

Medvedev, Zhores A. 1990. *The Legacy of Chernobyl*. New York: W.W. Norton.

Menzies, Heather. 1996. *Whose Brave New World? The Information Highway and the New Economy*. Toronto: Between The Lines.

Menzies, Heather. 1989. *Fastforward and Out of Control: How Technology is Changing Your Life*. Toronto: Macmillan.

Merchant, Carolyn. 1980. *The Death of Nature*. New York: Harper & Row.

Mesthene, Emmanuel G. 1970. *Technological Change: Its Impact on Man and Society*. Cambridge, Mass.: Harvard University Press.

Meyrowitz, Joshua. 1985. *No Sense of Place: The Impact of Electronic Media on Social Behavior.* Oxford: Oxford University Press.

Midgley, Mary. 1992. *Science as Salvaton: A Modern Myth and its Meaning*. London: Routledge.

Miller, Leslie. 1996. "Girls' Preferences in Software Design: Insights from a Focus Group." *Interpersonal Computing and Technology* 4(2): 27-36.

Mitchell, Juliet and Ann Oakley (eds.). 1986. *What Is Feminism?* New York: Pantheon Books.

Mol, Hans. 1976. *Identity and the Sacred.* Agincourt, Canada: The Book Society of Canada.

Moursund, David. 1983. *School Administrators Introduction to Instructional Use of Computers.* Eugene, Oregon: International Council for Computers in Education.

Moursund, David. 1982. "Personal Computing for Elementary and Secondary School Students." In *Computer Literacy,* edited by Robert Seidel, Ronald Anderson, and Beverly Hunter. Orlando, Fl.: Academic Press.

Munger, G.F., and B. H, Loyd. 1989. "Gender and Attitudes Toward Computers and Calculators: Their Relationship to Math Performance." *Journal of Educational Computing Research* 5(2): 167-77.

"Nanofuture." 1990. Science and the Citizen. *Scientific American* 262(1) (January): 15-16.

National Association of Secondary School Principals [NASSP]. 1984. *High Tech Schools.* Reston, Virginia: NASSP.

National Center for Education Statistics. 1997. *Findings from The Condition of Education 1997: Women in Mathematics and Science* (NCES 97-982). Washington, D.C.: United States Department of Education. httb//nces.ed.gov/pubs97/9798.html.

National Research Council. 1994. *Women Scientists and Engineers Employed in Industry: Why So Few?* Washington, D.C.: National Academy Press.

Negley, Glenn and J. Max Patrick. 1952. *The Quest for Utopia.* New York: Henry Schuman.

Negroponte, Nicholas. 1995. *Being Digital.* New York: Alfred A. Knopf.

Nelkin, Dorothy. 1987. *Selling Science: How the Press Covers Science and Technology.* New York: W.H. Freeman.

Nesti, Arnaldo. 1990. "Implicit Religion: The Issues and Dynamics of a Phenomenon." *Social Compass* 37(4): 423-438.

Newsweek. 1979-1988. Vol. 93 to Vol. 112.

Nimkoff, M. 1957. "Obstacles to Innovation." In *Technology and Social Change*, edited by Francis Allen et al. New York: Appleton-Century-Crofts.

Noble, David F. 1992. *A World Without Women: The Christian Clerical Culture of Western Science.* New York: Alfred A. Knopf.

Noble, Douglas. 1991. *The Classroom Arsenal: Military Research, Information Technology, and Public Education.* London: Falmer Press.

Office of Technology Assessment [OTA]. 1982. *Informational Technology and Its Impact on American Education.* Washington, D.C.: OTA.

Ogburn, William. 1964. *On Culture and Social Change,* edited by O. Duncan. Chicago: University of Chicago Press.

Ogburn, William. 1957. "How Technology Causes Social Change." In *Technology and Social Change*, edited by Francis Allen et al. New York: Appleton-Century-Crofts.

Ogburn, William. 1950. *Social Change.* New York: Viking Press.

Ogburn, William. 1946. *The Social Effects of Aviation.* Boston: Houghton Mifflin .

Ogden, Frank. 1995. *Navigating in Cyberspace: A Guide to the Next Millennium.* Toronto: Macfarlane Walter and Ross.

O'Keefe, Daniel. 1982. *Stolen Lightening: The Social Theory of Magic.* New York: Continuum.

Organization for Economic Co-operation and Development (OECD). 1983. *Assessing the Impacts of Technology on Society.* Paris: OECD.

Pacey, Arnold. 1990. *Technology in World Civilization.* Cambridge, Mass.: MIT Press.

Pacey, Arnold. 1983. *The Culture of Technology.* Cambridge, Mass.: MIT Press.

Papert, Seymour A. 1980. *Mindstorms.* New York: Basic Books.

Parenti, Michael. 1993. *Inventing Reality: The Politics of the Mass Media.* New York: St. Martin's Press.

Pearl, A., M. Pollack, E. Riskin, B. Thomas, E. Wolf, and A. Wu. 1990. "Becoming a Computer Scientist." *Communications of the ACM* 33(11): 48-57.

Perry, R., and L. Greber. 1990. "Women and Computers: An Introduction." *Signs: Journal of Women in Culture and Society* 16(11): 74-101.

Petrovic´, Gajo. 1967. *Marx in the Mid-Twentieth Century.* Garden City, N.Y.: Doubleday.

Pinch, T. and W. Bijker. 1987. "The Social Construction of Facts and Artifacts: Or How the Sociology of Science and the Sociology of Technology Might Benefit Each Other." In *The Social Construction of Technological Systems,* edited by W. Bijker, T. Hughes, and T. Pinch Cambridge, Mass.: MIT Press

Pogrow, Stanley. 1983. *Education in the Computer Age.* Beverly Hills, Calif.: Sage Publications.

Porter, A., F. Rossini, S. Carpenter, A. Roper, R. Larson, and J. Tiller. 1980. *A Guidebook for Technology Assessment and Impact Analysis.* New York: North-Holland.

Propp, Vladimir. 1928/1968. *Morphology of the Folktale, second edition,* trans. by Lawrence Scott. Austin: University of Texas Press.

Restivo, Sal. 1995. "The Theory Landscape in Science Studies: Sociological Traditions." In *Handbook of Science and Technology Studies,* edited by S. Jasanoff, G. Markle, J. Petersen, and T. Pinch. Thousand Oaks: Sage Publications.

Ribner, Irving. 1966. "Preface." In *Christopher Marlowe's Doctor Faustus: Text and Major Criticism,* edited by Irving Ribner. New York: Odyssey Press.

Ricoeur, Paul. 1976. *Interpretation Theory: Discourse and the Surplus of Meaning.* Fort Worth: Texas Christian University Press.

Ricoeur, Paul. 1970. *Freud and Philosophy*, trans. by D. Savage. New Haven: Yale University Press.

Ricouer, Paul. 1967. *The Symbolism of Evil,* trans. by Emerson Buchanan. Boston: Beacon Press.

Rosenthal, N. R., and D. M. Demetrulias. 1988. "Assessing Gender Bias in Computer Software." *Computers in the Schools* 5(1/2): 153-63.

Roszak, Theodore. 1986. *The Cult of Information.* New York: Pantheon Books.

Ruether, Rosemary Radford. 1972. *Liberation Theology.* New York: Paulist Press.

Russell, Jeffrey Burton. 1986. *Mephistopheles: The Devil in the Modern Mind.* Ithaca, N.Y.: Cornell University Press.

Saul, John Ralston. 1995. *The Unconscious Civilization.* Concord, Ont.: Anansi.

Schiebinger, Londa. 1989. *The Mind Has No Sex? Women in the Origins of Modern Science.* Cambridge, Mass.: Harvard University Press.

Schon, D. 1963. "Champions for Radical New Inventions." *Harvard Business Review* (March/April): 77-86.

Schutz, Alfred. 1973. *Collected Papers, Vol. I. The Problem of Social Reality,* edited by M. Natanson. The Hague: Martinus Nijoff.

Searle, John R. 1990. "Is The Brain's Mind A Computer Program?" *Scientific American* 262(1): 26-31.

Segal, Howard P. 1986. "The Technological Utopians." In *Imagining Tomorrow: History, Technology, and the American Future,* edited by Joseph Corn. Cambridge, Mass.: MIT Press.

Seidel, Robert J. 1982. "On the Development of an Information Handling Curriculum: Computer Literacy, A Dynamic Concept." In *Computer Literacy,* edited by Robert Seidel, Ronald Anderson, and Beverly Hunter. Orlando, Fl.: Academic Press.

Seligmann, Kurt. 1948. *Magic, Supernaturalism and Religion.* New York: Pantheon Books.

Shcherbak, Yuri M. 1996. "Ten Years of the Chornobyl Era." *Scientific American* 274(4): 44-49.

Sims, G. 1990. *The Anti-Nuclear Game.* Ottawa: University of Ottawa Press.

Smith, Dorothy. 1990. *The Conceptual Practices of Power.* Toronto: University of Toronto Press.

Smith, Dorothy. 1987. *The Everyday World as Problematic.* Boston: Northeastern University Press.

Smith, Michael L. 1983. "Selling the Moon: The U.S. Manned Space Program and the Triumph of Commodity Scientism." In *The Culture of Consumption,* edited by Richard Fox and T. J. Jackson Lears. New York: Pantheon Books.

Speer, Albert. 1970. *Inside the Third Reich,* trans. by Richard and Clara Winston. New York: Avon Books.

Spender, Dale. 1995. *Nattering on the Net: Women, Power and Cyberspace.* Toronto: Garamond Press.

Stahl, William. 1994. "Computer Literacy and Beyond." In *Future of Computerization in Institutions of Higher Learning,* edited by Binod Agrawal and Larry Symes. New Delhi: Concept Publishing.

Stark, Rodney and William Bainbridge. 1985. *The Future of Religion.* Berkeley: University of California Press.

Statistics Canada. 1996. *Education in Canada, 1996.* Ottawa: Minister of Supply and Services.

Statistics Canada. 1992. *Education in Canada: A Statistical Review for 1990-91.* Ottawa: Minister of Supply and Services.

Statistics Canada. 1986. *Education in Canada: A Statistical Review for 1984-85.* Ottawa: Minister of Supply and Services.

Staudenmaier, John M. S. J. 1994. "Rationality versus Contingency in the History of Technology." In *Does Technology Drive History? The Dilemma of Technological Determinism*, edited by M. R. Smith and L. Marx. Cambridge, Mass: MIT Press: 259-73.

Staudenmaier, John M. S. J. 1988. "Advent for Capitalists: Grief, Joy, and Gender in Contemporary Society." *The Nash Lecture* Nov. 1987. Regina, Sask.: Campion College, University of Regina.

Staudenmaier, John M. S. J. 1985. *Technology's Storytellers.* Cambridge, Mass: MIT Press.

Stix, Gary. 1996. "Waiting for Breakthroughs." *Scientific American* 274(4): 94-99.

Stoll, Clifford. 1995. *Silicon Snake Oil: Second Thoughts on the Information Highway.* New York: Doubleday.

Strauss, Stephen. 1988. " The Race For Space." *The Globe and Mail* 30 January, D1 and D4.

Suchman, L., and B. Jordan 1989. "Computerization and Women's Knowledge." In *Women, Work and Computerization: Forming New Alliances,* edited by K. Tijdens, M. Jennings, I.Wagner, and M. Weggelaar. Amsterdam: North-Holland.

Sullivan G. and P. O'Conner. 1988. "Women's Role Portrayals in Magazine Advertising: 1958-1983." *Sex Roles* 18 (3/4): 181-88.

Taylor, F. Sherwood. 1974. *The Alchemists.* New York: Arno Press.

Technorealism. 1998. *Principles of Technorealism.* www. technorealism.org.

Tenner, Edward. 1996. *Why Things Bite Back: Technology and the Revenge of Unintended Consequences.* New York: Alfred A. Knopf.

ter Borg, Meerten. 1992. "Mythology in Modern Society: Richartd Wagner and Bob Dylan." Paper presented at the annual meetings of the Society for the Scientific Study of Religion, Washington, D.C.

Thomas, J. Mark. 1987. *Ethics and Technoculture.* Lanham, Md: University Press of America.

Tillich, Paul. 1971. *Political Expectations,* edited by James Luther Adams. New York: Harper & Row

Time. 1979-1988. Vol. 113 to Vol. 132.

Toffler, Alvin. 1970. *Future Shock.* New York: Random House.

Tong, Rosemarie. 1989. *Feminist Thought.* Boulder: Westview Press.

Tuchman, Gaye. 1978. *Making News: A Study in the Construction of Reality.* New York: Free Press.

Turkle, Sherry. 1995. *Life on the Screen: Identity in the Age of the Internet.* New York: Simon and Schuster.

Turkle, Sherry. 1984a. *The Second Self: Computers and the Human Spirit.* New York: Simon and Schuster.

Turkle, Sherry. 1984b. "Women and Computer Programming: A Different Approach." *Technology Review* 87(Nov-Dec): 48-50.

Turkle, S., and S. Papert. 1990. "Epistemological Pluralism: Styles and Voices Within the Computer Culture." *Signs: Journal of Women in Culture and Science* 16(11): 128-57.

Turner, J. 1984. "Why More Women Than Men Shun Computers." *Chronicle of Higher Education* 29(Oct 10), 26ff.

Valenza, Joyce. 1997. "Girls + Technology = Turnoff?" *Technology Connection* 3(10): 20-21.

Vaughan, Dianne. 1996. *The Challenger Launch Decision: Risky Technology, Culture, and Deviance at NASA.* Chicago: University of Chicago Press.

Verne, G. 1986. "Women's Challenge to Computer Science and Technology." In *Women Challenge Technology. European Conference On Women, Natural Sciences and Technology*, edited by M. Dahms et al. Elsinore, Denmark. University of Aalborg: Centertrykkeriet.

Vico, Giambattista. 1744/1970. *The New Science,* Abridged and trans. by T. Bergin and M. Fisch. Ithaca, N.Y.: Cornell University Press.

Wach, Joachim. 1944. *Sociology of Religion.* Chicago: University of Chicago Press.

Watt, Daniel H. 1982. "Education for Citizenship in a Computer-Based Society." In *Computer Literacy,* edited by Robert Seidel, Ronald Anderson, and Beverly Hunter. Orlando, Fl.: Academic Press.

Wax, Rosalie and Murray Wax. 1962. "The Magical World View." *Journal for the Scientific Study of Religion* 1(2): 179-88.

Weart, Spencer. 1988. *Nuclear Fear: A History of Images.* Cambridge, Mass.: Harvard University Press.

Weber, Max. 1949. *The Methodology of the Social Sciences*, trans. and edited by Edward Shils and Henry Finch. New York: Free Press.

Weber, Max. 1947. *The Theory of Social and Economic Organization,* trans. by A.M. Henderson and Talcott Parsons. New York: Free Press.

Weber, Max. 1922/1963. *The Sociology of Religion,* trans. by E. Fischoff, intro. by T. Parsons. Boston: Beacon Press.

Weber, Max. 1905/1958. *The Protestant Ethic and the Spirit of Capitalism,* trans. by Talcott Parsons. New York: Charles Scribner's Sons.

Webster, Hutton. 1948. *Magic: A Sociological Study.* New York: Octagon Books.

Weinberg, Alvin M. 1966/1986. "Can Technology Replace Social Engineering?" In *Technology and the Future,* edited by Albert H.Teich. New York: St. Martin's Press.

Weizenbaum, Joseph. 1985. "The Myths of Artificial Intelligence." In *The Information Technology Revolution,* edited by Tom Forester. Cambridge, Mass.: MIT Press.

Westrum, Ron. 1991. *Technologies and Society: The Shaping of People and Things.* Belmont, Calif.: Wadsworth.

Whitehead, Alfred North. 1933. *Adventures of Ideas.* New York: Free Press.

Whitehead, Alfred North. 1929 *Process and Reality.* New York: Free Press.

Wiener, Lauren Ruth. 1993. *Digital Woes: Why We Should Not Depend on Software.* Reading, Mass.: Addison-Wesley Publishing Company.

Winner, Langdon. 1993. "Upon Opening the Black Box and Finding It Empty: Social Constructivism and the Philosophy of Technology." *Science, Technology and Human Values* 18(3): 362-78.

Winner, Langdon. 1985. "Do Artifacts Have Politics?" In *The Social Shaping of Technology,* edited by D. MacKenzie and J. Wajcman. Milton Keynes, England: Open University Press.

Wise, John. 1996. "The Modem Is No Match For The Sword." *Time* August 19, 1996: 56.

Wise, Patricia. 1997. "Always Already Virtual: Feminist Politics in Cyberspace." In *Virtual Politics: Identity and Community in Cyberspace,* edited by David Holmes. Thousand Oaks, Calif.: Sage Publications.

Woolgar, Steve. 1991. "The Turn To Technology In Social Studies Of Science." *Science, Technology and Human Values* 16(1): 20-50.

Yelland, Nicola. 1995. "Young Children's Attitudes to Computers and Computing." *Australian Journal of Early Childhood* 20(2): 20-25.

Yeloushan, K. 1989. "Social Barriers Hindering Successful Entry of Females into Technology-Oriented Fields." *Educational Technology* 29(11): 44-46.

Ziguras, Christopher. 1997. "The Technologization of the Sacred: Virtual Reality and the New Age." In *Virtual Politics: Identity and Community in Cyberspace,* edited by David Holmes. Thousand Oaks, Calif.: Sage Publications.

Index

Series Published by Wilfrid Laurier University Press for the Canadian Corporation for Studies in Religion / Corporation Canadienne des Sciences Religieuses

Editions SR

1. *La langue de Ya'udi : description et classement de l'ancien parler de Zencircli dans le cadre des langues sémitiques du nord-ouest*
   Paul-Eugène Dion, O.P. / 1974 / viii + 511 p. / OUT OF PRINT
2. *The Conception of Punishment in Early Indian Literature*
   Terence P. Day / 1982 / iv + 328 pp.
3. *Traditions in Contact and Change: Selected Proceedings of the XIVth Congress of the International Association for the History of Religions*
   Edited by Peter Slater and Donald Wiebe with Maurice Boutin and Harold Coward
   1983 / x + 758 pp. / OUT OF PRINT
4. *Le messianisme de Louis Riel*
   Gilles Martel / 1984 / xviii + 483 p.
5. *Mythologies and Philosophies of Salvation in the Theistic Traditions of India*
   Klaus K. Klostermaier / 1984 / xvi + 549 pp. / OUT OF PRINT
6. *Averroes' Doctrine of Immortality: A Matter of Controversy*
   Ovey N. Mohammed / 1984 / vi + 202 pp. / OUT OF PRINT
7. *L'étude des religions dans les écoles : l'expérience américaine, anglaise et canadienne*
   Fernand Ouellet / 1985 / xvi + 666 p.
8. *Of God and Maxim Guns: Presbyterianism in Nigeria, 1846-1966*
   Geoffrey Johnston / 1988 / iv + 322 pp.
9. *A Victorian Missionary and Canadian Indian Policy: Cultural Synthesis vs Cultural Replacement*
   David A. Nock / 1988 / x + 194 pp. / OUT OF PRINT
10. *Prometheus Rebound: The Irony of Atheism*
    Joseph C. McLelland / 1988 / xvi + 366 pp.
11. *Competition in Religious Life*
    Jay Newman / 1989 / viii + 237 pp.
12. *The Huguenots and French Opinion, 1685-1787: The Enlightenment Debate on Toleration*
    Geoffrey Adams / 1991 / xiv + 335 pp.
13. *Religion in History: The Word, the Idea, the Reality / La religion dans l'histoire : le mot, l'idée, la réalité*
    Edited by/Sous la direction de Michel Despland and/et Gérard Vallée
    1992 / x + 252 pp.
14. *Sharing Without Reckoning: Imperfect Right and the Norms of Reciprocity*
    Millard Schumaker / 1992 / xiv + 112 pp.
15. *Love and the Soul: Psychological Interpretations of the Eros and Psyche Myth*
    James Gollnick / 1992 / viii + 174 pp.
16. *The Promise of Critical Theology: Essays in Honour of Charles Davis*
    Edited by Marc P. Lalonde / 1995 / xii + 146 pp.

17. *The Five Aggregates: Understanding Theravāda Psychology and Soteriology*
Mathieu Boisvert / 1995 / xii + 166 pp.
18. *Mysticism and Vocation*
James R. Horne / 1996 / vi + 110 pp.
19. *Memory and Hope: Strands of Canadian Baptist History*
Edited by David T. Priestley / 1996 / viii + 211 pp.
20. *The Concept of Equity in Calvin's Ethics**
Guenther H. Haas / 1997 / xii + 205 pp.
*** Available in the United Kingdom and Europe from Paternoster Press.**
21. *The Call of Conscience: French Protestant Responses to the Algerian War, 1954-1962*
Geoffrey Adams / 1998 / xxii + 270 pp.
22. *Clinical Pastoral Supervision and the Theology of Charles Gerkin*
Thomas St. James O'Connor / 1998 / x + 152 pp.
23. *Faith and Fiction: A Theological Critique of the Narrative Strategies of Hugh MacLennan and Morley Callaghan*
Barbara Pell / 1998 / v + 141 pp.
24. *God and the Chip: Religion and the Culture of Technology*
William A. Stahl / 1999 / vi + 186 pp.

Comparative Ethics Series / Collection d'Éthique Comparée

1. *Muslim Ethics and Modernity: A Comparative Study of the Ethical Thought of Sayyid Ahmad Khan and Mawlana Mawdudi*
Sheila McDonough / 1984 / x + 130 pp. / OUT OF PRINT
2. *Methodist Education in Peru: Social Gospel, Politics, and American Ideological and Economic Penetration, 1888-1930*
Rosa del Carmen Bruno-Jofré / 1988 / xiv + 223 pp.
3. *Prophets, Pastors and Public Choices: Canadian Churches and the Mackenzie Valley Pipeline Debate*
Roger Hutchinson / 1992 / xiv + 142 pp. / OUT OF PRINT
4. *In Good Faith: Canadian Churches Against Apartheid*
Renate Pratt / 1997 / xii + 366 pp.

Dissertations SR

1. *The Social Setting of the Ministry as Reflected in the Writings of Hermas, Clement and Ignatius*
Harry O. Maier / 1991 / viii + 230 pp. / OUT OF PRINT
2. *Literature as Pulpit: The Christian Social Activism of Nellie L. McClung*
Randi R. Warne / 1993 / viii + 236 pp.

Studies in Christianity and Judaism / Études sur le christianisme et le judaïsme

1. *A Study in Anti-Gnostic Polemics: Irenaeus, Hippolytus, and Epiphanius*
Gérard Vallée / 1981 / xii + 114 pp. / OUT OF PRINT
2. *Anti-Judaism in Early Christianity*
Vol. 1, *Paul and the Gospels*
Edited by Peter Richardson with David Granskou / 1986 / x + 232 pp.
Vol. 2, *Separation and Polemic*
Edited by Stephen G. Wilson / 1986 / xii + 185 pp.